Industry and Environment in Latin America

Most discussion of environmental issues in Latin America has focused on the 'green issues' of deforestation and land degradation. Although the majority of Latin Americans live in urban areas and the region has industrialised and urbanised rapidly, relatively little attention has been paid to the 'brown issues' such as industrial pollution. *Industry and Environment in Latin America* attempts to redress this imbalance.

Focusing on the three most industrialised countries in the region (Argentina, Brazil and Mexico), the book situates the environmental impact of industry in the context of the recent liberalisation of the Latin American economies. In addition to providing regional overviews of the environmental impact of the manufacturing sector and mining, the book includes detailed case studies of some of the most polluting industries, and evaluates the development of environmental management and pollution prevention. The conclusion emerging from these studies is that Latin American industry is still a long way from sustainability.

Offering a wealth of empirical research material, and an important new perspective on the environmental implications of globalisation, this book represents essential reading for any scholar of environment or development-related subjects.

Rhys Jenkins is a Reader at the School of Development Studies, University of East Anglia, Norwich. He has worked extensively on issues of industrial development in Latin America, and on the impact of globalization. His previous publications include *Transnational Corporations and Industrial Transformation in Latin America* and *Transnational Corporations and the Latin American Motor Vehicle Industry*.

Routledge Research Global Environmental Change Series

1. **The Condition of Sustainability**
   Ian Drummond and Terry Marsden
2. **Working the Sahel**
   Environment and society in northern Nigeria
   Michael Mortimore and Bill Adams
3. **Global Trade and Global Social Issues**
   Edited by Annie Tyalor and Caroline Thomas
4. **Environmental Policies and NGO Influence**
   Land degradation and sustainable resource management in sub-Saharan Africa
   Edited by Alan Thomas, Susan Carr and David Humphreys
5. **The Sociology of Energy, Buildings and the Environment**
   Constructing knowledge, designing practice
   Simon Guy and Elizabeth Shove
6. **Living with Environmental Change**
   Social vulnerability, adaptation and resilience in Vietnam
   Edited by W. Neil Adger, P. Mick Kelly and Nguyen Huu Ninh
7. **Land and Limits**
   Interpreting sustainability in the planning process
   Susan Owens and Richard Cowell
8. **The Business of Greening**
   Edited by Stephen Fineman
9. **Industry and Environment in Latin America**
   Edited by Rhys Jenkins

Also available in the Routledge Global Environmental Change series:

Timescapes of Modernity
Barbara Adam
Reframing Deforestation
James Fairhead and Melissa Leach
British Environmental Policy and Europe
Edited by Philip Lowe
The Politics of Sustainable Development
Edited by Susan Baker, Maria Kousis, Dick Richardson and Stephen Young
Argument in the Greenhouse
Mick Mabey, Stephen Hall, Clare Smith and Sujata Gupta

Environmentalism and the Mass Media
Graham Chapman, Keval Kumar, Caroline Fraser and Ivor Gaber
Environmental Change in Southeast Asia
Edited by Michael Parnwell and Raymond Bryant
The Politics of Climate Change
Edited by Timothy O'Riordan and Jill Jagger
Population and Food
Tim Dyson
The Environment and International Relations
Edited by John Vogler and Mark Imber
Global Warming and Energy Demand
Edited by Terry Barker, Paul Ekins and Nick Johnstone
Social Theory and the Global Environment
Michael Redclift and Ted Benton

Industry and Environment in Latin America

Edited by Rhys Jenkins

London and New York

First published 2000
by Routledge
2 Park Square, Milton Park, Abingdon, Oxon, OX14 4RN

Simultaneously published in the USA and Canada
by Routledge
270 Madison Ave, New York NY 10016

Routledge is an imprint of the Taylor & Francis Group

Transferred to Digital Printing 2006

Typeset in Baskerville by Keyword Typesetting Services Ltd

British Library Cataloguing in Publication Data
A catalogue record for this book is available from the British Library

Library of Congress Cataloging-in-Publication Data
Industry and environment in Latin America/edited b Rhys Jenkins.
p. cm.
Includes bibliographical references and index.
ISBN 0-415-23447-6 (alk. paper)
1. Industries—Environmental aspects—Latin America. 2. Industries—Environmental aspects—Brazil. 3. Industries–Environmental aspects—Mexico. I. Jenkins, Rhys.

HC130.E5 I53 2000 00-057634
333.7′098–dc21

ISBN 0-415-23447-6

CONTENTS

FIGURES

TABLES

CONTRIBUTORS

Jonathan Barton is a lecturer in Environmental Policy at the School of Development Studies, UEA. His research focuses on business environmental management, environmental regulations and the globalisation of the environment industry.

Flor Brown is currently a lecturer at the Graduate Faculty of Economics of the National Autonomous University of Mexico (UNAM). She has published on industrial productivity, technical change and industrial linkages and networks in Mexican industry.

Daniel Chudnovsky is Director of the Centro de Investigaciones para la Transformación (CENIT) in Buenos Aires and is professor of Development Economics at the University of Buenos Aires. He has written extensively on trade, industrial restructuring, foreign direct investment, environmental and technology issues, mostly in relation to Latin America.

Sonia Maria Dalcomuni is a lecturer and researcher at the Department of Economics at Universidade Federal do Espírito Santo. She works on technology, environment and development. Currently, she is the co-ordinator of the Postgraduate Programme on Economics and a Research Group Leader in this field.

Lilia Domínguez-Villalobos is professor in industrial economics at the Faculty of Economics of the National Autonomous University of Mexico (UNAM). She has researched on industrial organisation, new technologies and productivity in Mexican manufacturing, and entrepreneurial behaviour in relation to pollution abatement and the possibility of green innovations.

Valeria Freylejer is an economist and research assistant at the Centro de Investigaciones para la Transformación (CENIT) in Buenos Aires.

Nia Hughes-Witcomb currently works for Billiton in their Corporate HSE Department in Johannesburg. She was previously a research fellow in the Mining and Energy Research Network, supported partly by the Department for International Development, Environment Research programme, and has worked as a lecturer in the UK and China.

Rhys Jenkins is Reader in Economics at the School of Development Studies, UEA. He has researched and written extensively on industrial development in Latin America, the impacts of globalisation on developing countries, and the links between trade, investment and the environment.

Andrés Lopez is a researcher at the Centro de Investigaciones para la Transformación (CENIT) in Buenos Aires and professor in graduate as well as in undergraduate courses at the University of Buenos Aires. His main research work has been done in areas such as foreign direct investment, industrial organisation and environmental management.

Alfonso Mercado is a Research Fellow at El Colegio de México, and Chair of the Department of Economic Studies at El Colegio de la Frontera Norte in Tijuana, B.C., Mexico. His recent publications are related to environmental economics, industrial organisation and labour markets.

Jan Thomas H. Odegard has an M. Phil. in Development Geography from the University of Oslo, Norway. He has done several years of research on the industrial development of Brazil, and has worked extensively with NGOs on issues of development and the environment. Currently, he works for UNIDO in Mozambique.

Alyson Warhurst is Professor of Corporate Strategy and International Development at Warwick Business School and also Co-Director of the Corporate Citizenship Unit based at the Centre for Creativity, Strategy and Change. She continues to direct the international Mining and Energy Research Network (MERN), which she established in 1991.

PREFACE

Preliminary versions of all the contributions to this book, with the exception of Chapter 3, were originally presented at two panels which formed part of the Environment track at the XXI International Congress of the Latin American Studies Association in Chicago in September 1998. The Economic and Social Research Council's Global Environmental Change Programme (GEC) provided financial support for the travel and accommodation of several of the participants. The three chapters on Mexico and parts of Chapter 2 are based on work carried out as part of a GEC-funded research project on *Pollution, Trade and Investment: Case Studies of Mexico and Malaysia* (Award Number L320253248), while Chapter 3 draws on the GEC project *Technology Transfer and the Diffusion of Clean Technology*. The research reported in Chapters 5 and 6 on Brazil was done under the auspices of a European Union funded project entitled *Environmental Regulation, Globalisation of Production and Technological Change* (Contract No. ENV4-CT96-0235).

1

INTRODUCTION

Rhys Jenkins

The conventional view of environmental problems in Latin America is dominated by the so-called 'green issues'. In the media the problems of deforestation and the protection of the rainforest dominate discussion of environmental issues in the region. This is also true of academic fora. At the 1998 international congress of the Latin American Studies Association (LASA), where the initial versions of most of the papers included in this volume were presented, almost a third of the 28 panels in the environmental track were exclusively devoted to forest-related issues. Other issues featuring prominently at the congress included biodiversity, resource management and eco-tourism.

It is not surprising that the Northern media and US academics give so much attention to these issues since global warming poses a threat to the North as well as the South, and biodiversity is also a global issue. In contrast, relatively little attention is paid to the 'brown issues' of urban and industrial pollution. Apart from the papers in this volume, only one other panel at the LASA congress was devoted to industrial and urban pollution.

This is paradoxical since three-quarters of the population of Latin America live in urban areas, and industrial production is more significant in the region than agriculture, fisheries and forestry combined.[1] However, industrial pollution continues to be seen mainly as a problem of Northern industrial economies and to have little relevance to the South. Nevertheless, although Latin America accounts for only 6 per cent of world manufacturing value added, certain cities and industrial corridors in the region are highly industrialised and as a result suffer from major pollution problems. Mining in the region has also had a major environmental impact over many years.

Although not reflected in the academic literature on environmental problems in Latin America, anyone who has spent any time in the major cities of the region cannot but be aware of the severity of urban environmental problems including high levels of atmospheric pollution, contamination of water supplies and problems of waste disposal including, critically, hazardous industrial waste.

Data on air pollution indicates that, in a number of Latin American cities, concentrations of pollutants are above the World Health Organization

Table 1.1 Air pollution levels[a] in selected cities *c.* 1995

	TSP	SO_2	NO_2
Cordoba (Arg)	97	N/A	97
São Paulo (Br)	86	43	83
Rio de Janeiro (Br)	139	129	N/A
Santiago (Chile)	N/A	29	81
Bogota (Col)	120	N/A	N/A
Guayaquil (Ec)	127	15	N/A
Quito (Ec)	175	31	N/A
Mexico City (Mex)	279	74	130
Guadalajara (Mex)	156	40	174
Monterrey (Mex)	46	23	112
Caracas (Ven)	53	33	57
WHO Guidelines	90	50	50

Sources: World Bank (1999) *World Development Indicators*: Table 3.13; OECD (1999) *Environmental Data Compendium, 1999*: Table 2.4.

Note

a In micrograms per cubic metre.

(WHO) guidelines for air quality standards. None of the major Latin American cities for which data are available falls within the guidelines for all the three pollutants covered (see Table 1.1).

Although industry is by no means the only factor contributing to air pollution in urban areas it is, together with vehicle emissions, a major element. In São Paulo, for instance, industry accounted for 88 per cent of sulphur dioxide emissions, 65 per cent of particulates and 24 per cent of nitrogen oxides (Shaman, 1996: 6). In Mexico City the corresponding figures are 70 per cent for sulphur dioxide, 24 per cent for particulates and 25 per cent for nitrogen oxides, while in Monterrey they come to 92 per cent, 89 per cent and 36 per cent, respectively[2] (Poder Ejecutivo Federal, 1996: 79; INE, 1997: Table 20).

There are similar problems with water quality with respect to the regions' rivers, lakes and coastal waters. In Mexico, in the early 1990s, 20 out of 29 main watersheds for which there was information were classified as excessively or strongly polluted (USAID, 1995: 1–10). In Brazil stretches of the Paraiba do Sul river in the state of Rio de Janeiro and several of its tributaries are badly polluted, as is Guanabara Bay (World Bank, 1996). In March 2000 a popular lagoon in Rio was contaminated by a leaking sewer, which killed millions of fish. In Buenos Aires 20,000 plants (nearly two-thirds of them with no treatment facilities) discharge into the Rio Matanza-Riachuelo, one of the most polluted rivers in the metropolitan area (World Bank, 1995).

This book sets out to fill the major gap that exists in the literature on environmental problems in Latin America. It brings together a number of detailed case studies of industrial pollution in Argentina, Brazil and Mexico and represents a first attempt to explore some of the issues raised by the environmental impacts of industry in the region.

Trends in industrial pollution in Latin America

It is difficult to obtain estimates of industrial emissions and effluent in Latin America, particularly over time. This partly reflects the lack of monitoring of pollution in the past in the region. For example, although a number of inventories of emissions have been carried out in Mexico City since the early 1980s, these are not really comparable from one year to another since they use different US Environmental Protection Agency (EPA) conversion factors to estimate the pollution load in different years.

One type of emission for which international estimates are available over time is carbon dioxide. Figure 1.1 shows the increase in carbon dioxide emissions in Argentina, Brazil and Mexico since 1970. This shows a clear pattern in Brazil and Mexico with substantial growth in the 1970s, a drop in the early 1980s as the economic crisis hit the region and further growth from the mid-1980s. Argentina shows something of the same pattern but in a much less marked fashion.

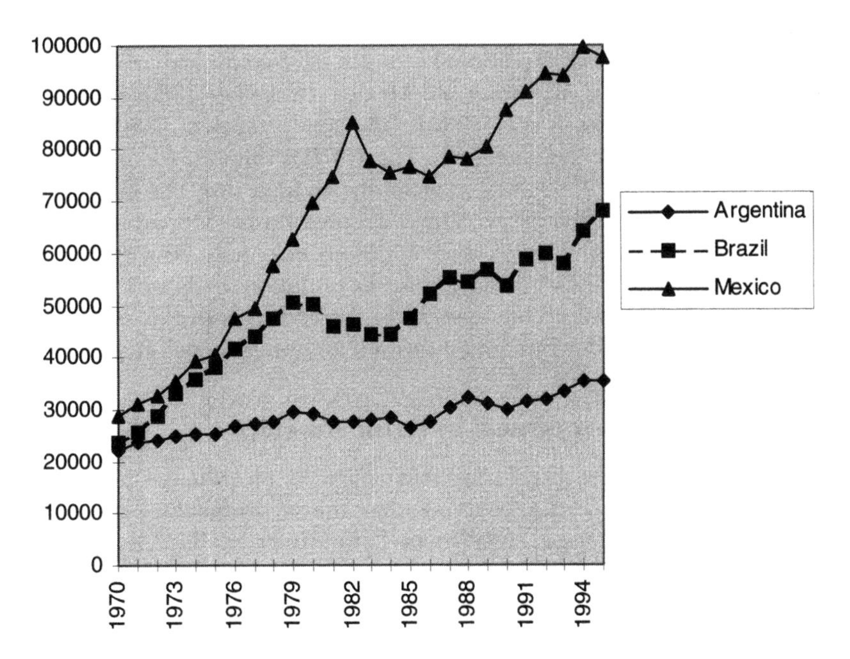

Figure 1.1 Carbon dioxide emissions 1970–95 (thousand metric tonnes of carbon). Source: Gregg Marland and Tom Boden (Oak Ridge National Laboratory).

Unfortunately, similar estimates are not available for other industrial emissions for the three countries. Therefore, in order to obtain countrywide estimates of industrial emissions for a much wider range of pollutants another set of indicators were calculated using the World Bank's Industrial Pollution Projection System (IPPS). The IPPS consists of a set of coefficients which relate emissions of pollutants to value added, output or employment. These have been calculated by the World Bank from US data on emissions and industrial production. These coefficients can then be applied to industrial data for other countries to obtain pollution estimates.

In order to estimate emissions associated with industrial production in Argentina, Brazil and Mexico, United Nations Industrial Development Organization (UNIDO) data on industrial value added for the three countries were converted into 1987 US dollars. Because the IPPS coefficients derive from the USA, which has stricter environmental regulations than are generally found in Latin America, it is likely that they underestimate the level of emissions in the region. On the other hand, because fixed coefficients from 1987 were used, no account is taken of technological improvements which reduce emissions per dollar of value added and they will therefore tend to exaggerate the growth of pollution.

The growth of industrial emissions in Argentina, Brazil and Mexico from 1975 to 1995 is depicted in Fig. 1.2. The data represented include four major air pollutants, one water pollutant and toxic emissions. The pattern which they show is slightly different from that observed in the previous figure for carbon dioxide. The growth of pollution seems to have continued up until the mid-1980s, particularly in Argentina and Mexico. In the late 1980s emissions fell only to increase again in the 1990s in all three countries. These figures need to be treated with some caution. The UNIDO figures for industrial growth in Brazil in the 1990s are suspiciously high so that the growth in emissions in the 1990s have been adjusted downwards. Moreover, since, unlike the data on carbon dioxide, these emissions have only been estimated at five-year intervals, the exact turning points in pollution levels are not clear. What is consistently apparent, however, is the fact that renewed growth after the 'lost decade' of the 1980s has brought with it an upturn in pollution levels.

'Brown issues' in Latin America

Although there is a lack of direct empirical evidence, the estimates presented above make a strong case for believing that the environmental effects of industrialisation is a growing problem in Latin America. This leads to a number of questions that need to be considered.

First, to what extent are these problems being addressed by governments and by business in the region? Is the general neglect of such issues in the academic literature paralleled by a low priority given to industrial pollution control on the part of Latin American governments and firms operating in

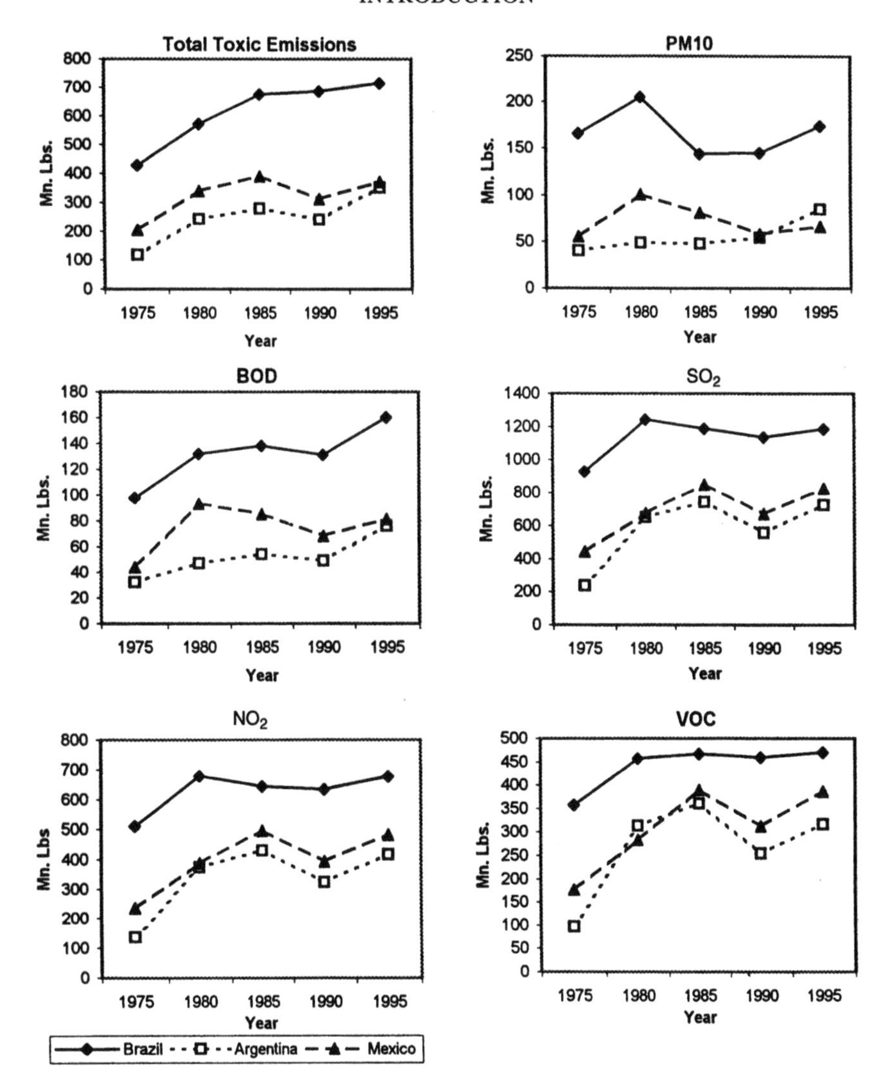

Figure 1.2 Emissions of selected pollutants in Argentina, Brazil and Mexico.

the region? Second, in so far as business is attempting to mitigate the environmental problems which it creates, what are the main drivers behind such efforts and what are the major obstacles which firms face? Third, how have the recent economic changes in Latin America, involving greater integration with the world economy, deregulation of domestic markets and privatisation of state enterprises, affected environmental performance in the region?

Environmental issues are a relatively recent addition to the public policy agenda in Latin America. Major institutional and policy changes which increased the prominence of these issues date from the late 1980s or early 1990s in Argentina, Brazil and Mexico. In 1988 Brazil's new constitution reorganised the country's environmental system, creating a Secretariat for the Environment (SENAM) as a central agency and the Institute of the Environment and Renewable Natural Resources (IBAMA) with responsibility for monitoring and enforcement. In the same year, the Mexican government passed the General Law for Ecological Equilibrium and Environmental Protection. In 1991 Argentina created a Secretariat of Natural Resources and Human Environment within the Office of the President and a year later passed a hazardous waste law. Also, in 1992 Mexico set up the National Ecology Institute (INE) and the Federal Attorney's Office for Environmental Protection (PROFEPA). Two years later, both Brazil and Mexico further reorganised their environmental systems with the creation of the Ministry of Environment, Water Resources and the Legal Amazon in Brazil, and the Secretariat of Environment, Natural Resources and Fisheries (SEMARNAP) in Mexico.

Despite these changes, there is often a perceived conflict between environmental protection and economic policy objectives such as increasing exports or attracting direct foreign investment. Environmental agencies or ministries are usually politically weak, particularly in relation to economic ministries. Regulatory bodies are often understaffed and unable to enforce fully regulations and standards.

Similarly, firms have until recently shown very little concern about their environmental impact in Latin America, with disastrous consequences. The city of Cubataõ in Brazil was in the 1970s a notorious example of the environmental consequences of unregulated industrial development. Many parts of the Mexican border with the USA which have experienced rapid industrial growth are today paying the environmental price. Major industrial accidents such as the explosions in the San Juanico district of Mexico City in 1984 and Guadalajara in 1992 are further illustrations of the dangers of inadequate regulation and a lack of concern on the part of firms.

There is some evidence that this situation is also gradually changing as firms adopt pollution abatement measures and begin to introduce environmental management systems. The environmental market in Latin America has been estimated to grow at 12 per cent per annum and to be worth a total of around $6.6 billion in 1995 (CEC, 1996: Table 1–1). This is still low, accounting for only 0.6 per cent of gross domestic product (GDP) compared to between 2 and 3 per cent in Europe and the USA, but represents an improvement on the position in the 1980s. There is also evidence of firms adopting environmental management systems. So far relatively few firms have obtained ISO 14000 certification in the region, but the total number is growing (see Table 1.2).

The papers in this book throw further light on the extent to which firms in particular industries in the major Latin American countries are addressing

Table 1.2 Number of firms with ISO 14000 in Latin America

	Dec. 1996	*Dec. 1997*	*June 1999*
Argentina	5	28	63
Brazil	6	63	88
Colombia	1	3	3
Costa Rica		2	3
Chile		2	5
Mexico	2	11	48
Peru			1

Source: Schaper (1999): Table IV.1.

environmental issues through adopting environmental policies, implementing environmental management systems and introducing pollution control equipment or adopting pollution prevention methods.

To the extent that some firms are beginning to change and to adopt more environmentally responsible attitudes, this raises the question of what are the main drivers which lead them to do so. Is this mainly a response to new government regulations and/or more effective enforcement by environmental regulators? Are there other factors which have led firms to adopt such measures? What has been the role of market pressures as a result of increased consumer awareness and NGO (non-government organisation) pressure on environmental issues? Are local community demands an important factor? What role do clients (particularly overseas) play in requiring better environmental performance? Have firms obtained economic advantages from adopting less polluting methods?

Related to this are questions about the obstacles which prevent some firms from adopting more environmentally friendly measures, and limit the extent of adoption by others. Are there conflicts between environmental and other objectives of the firm and what is the priority of top management? Is there a lack of incentive to adopt such measures? Is the technology available and do firms have sufficient information?

Another set of issues revolves around the response of different types of firms. There is some evidence that small and medium firms lag behind larger firms in adopting environmental measures. In Mexico, for instance, levels of particulate emissions per person employed are generally much higher in small firms than in medium or large ones (Dasgupta *et al.*, 1998). It has also been suggested that there is a strong relationship between the size of both firms and plants and managerial efforts in relation to the environment (Dasgupta *et al.*, 1997). Is size therefore a critical factor in determining environmental performance? It is also often argued that transnational corporations (TNCs) can play a major role in introducing high environmental standards in an industry. Do foreign subsidiaries in Latin America perform better than their local counterparts in the region?

The implications of globalisation for the environment have been a major area of debate in recent years. On the one hand it is argued that an increasingly deregulated global economy poses a threat to the environment. On the other hand there are those who predict a 'win–win' scenario in which increased economic efficiency as a result of globalisation is accompanied by environmental improvements, both directly through faster diffusion of environmental technologies and management systems, and indirectly through increased income levels, leading to greater public pressure for a better environment.

Latin America since the 1980s provides a good case study of these claims, since the region has undergone a major reorientation towards the world economy as a result of the widespread adoption of trade liberalisation and increased freedom of movement of capital. Has greater openness in the region contributed to improved environmental performance as firms try to meet the demands of more environmentally aware consumers in the North? Has increasing competition both in the export market and at home forced firms to concentrate on cutting costs to remain competitive and made it more difficult for them to internalise environmental costs? Has increased openness led to a more rapid diffusion of pollution abatement and cleaner production technologies? Have Latin American countries increasingly specialised in more polluting activities or has their comparative advantage been in relatively clean, labour-intensive industries? Has increased foreign investment helped to raise environmental standards in the region or has globalisation had a 'chilling effect' on environmental regulation?

Case studies

So far evidence on all these issues in Latin America is relatively limited and a major aim of this book is to provide more information to begin to answer some of these questions. It brings together studies by a number of Latin American and European authors who have been researching problems of industrial pollution in the region. The lack of previous research and the limited available data has led to an emphasis on detailed case studies of particular industries in the most industrialised Latin American countries. By their very nature, pollution problems are specific to particular industries, and the bulk of industrial pollution is generated by a relatively small number of industrial sectors such as chemicals, petrochemicals, ferrous and non-ferrous metals, non-metallic minerals, pulp and paper, and leather. Thus, a sectoral approach is particularly appropriate when discussing environmental impacts.

The next two chapters take a regional perspective, providing an overview of particular issues in manufacturing industry and mining, respectively. Chapter 2 looks at the impact of the recent economic liberalisation on industrial pollution in Argentina, Brazil and Mexico. It develops new estimates of pollution associated with exports and imports by the three countries in order to assess whether they are net exporters or importers of pollution, and how this has changed over time. In addition to looking at changes in the pattern of trade, it also discusses some of the other ways in which greater openness to both trade and investment could have an impact on pollution, examining the effect of liberalisation on the scale of production, and on the technology and management systems used within industries.

In Chapter 3, Alyson Warhurst and Nia Hughes-Witcomb consider the impact of mining in Latin America on the environment. Although the other papers in the volume focus on the manufacturing industry, mining is a sector which has major environmental impacts and, like manufacturing, has been relatively neglected in recent discussions of environmental issues in the region. Mining has also been affected in a major way by recent economic reforms in Latin America, particularly through privatisation and new mining codes and tax regimes. It has also been a major area of new foreign investment. It is, therefore, an excellent sector in which to consider the environmental implications of liberalisation.

The authors also consider the drivers which are leading firms to adopt less environmentally damaging practices in a number of Latin American countries. In addition to the more industrialised countries of the region which are the focus of the other chapters in the book, they draw on examples from Bolivia and Chile.

Chapter 4 is the only one to focus exclusively on Argentina. Unlike the other country case studies, it does not look at a specific industrial sector; however, it draws on a number of sectoral studies previously undertaken by the *Centro de Investigaciones para la Transformación* in the pulp and paper, petrochemicals, and iron and steel industries (Chudnovsky *et al.*, 1996). The chapter presents evidence from two surveys, one of large firms in Argentina and the other of small and medium enterprises in the Greater Buenos Aires area. It provides information both on the environmental management of these firms and on their adoption of pollution prevention practices. It considers the factors motivating firms to adopt pollution prevention and the obstacles which they face, and it looks at the performance of different types of firms.

The next three chapters all provide case studies of the environmental impact of particular industries in Brazil. Brazil accounts for around 40 per cent of the region's manufacturing value added and includes some of its most polluted industrial areas. The three industries studied are amongst the most polluting in the country.

Jonathan Barton presents the case of the iron and steel industry, concentrating particularly on the integrated producers which account for most of the pollution in the industry. The chapter illustrates the environmental advances made in the Brazilian industry since privatisation in the early 1990s and discusses the major factors which motivated these changes. He also carries out an analysis of the environmental performance of five of the major steel companies in Brazil.

Although not a heavy industry like steel, leather tanning is also a major polluter, particularly of water. In Chapter 6, Jan Thomas Odegard describes the restructuring which has taken place in the Brazilian leather industry in recent years, particularly the increased importance of exports of *wet blue* leather and the environmental implications of these changes. He also documents the differences in cost structure between the tanning industry in Brazil and Europe to highlight the significance of pollution abatement costs. Most of the firms involved in tanning are Small and Medium Enterprises (SMEs) and this may partly account for the lack of progress in environmental matters compared to some other industries.

The third Brazilian case study is of the market pulp industry. This is a particularly interesting industry in which to consider the environmental impact in Latin America because of the degree to which the industry has been a focus for environmental pressure groups in Europe. This has led to major innovations in the world pulp industry designed to eliminate the use of chlorine in the bleaching process. Sonia Dalcomuni examines the ways in which Brazilian pulp manufacturers have responded to these market pressures. A particular feature of this chapter is the focus on the dynamics of innovation in the Brazilian industry. This provides important insights into the drivers contributing to improved environmental performance at the firm level.

Mexico is the second largest industrial economy in Latin America, behind Brazil, and the next three chapters cover the environmental performance of three key industries. The links between globalisation and environmental protection have been particularly prominent in discussion of Mexico because of the debate at the time of the North American Free Trade Agreement (NAFTA) negotiations, and the subsequent conclusion of the environmental side-agreement and the creation of the North American Commission on Environmental Co-operation.

The textile industry is not usually thought of as being one of the most polluting branches of manufacturing. However, because of its size, it does make a significant contribution to total industrial pollution in Mexico, particularly in terms of water effluent where it is the fourth largest generator of pollution (INEGI, 1995: Table V.A.140). In Chapter 8, Flor Brown reports on a survey of Mexican textile firms and looks particularly at the way in which they have responded to trade liberalisation. The paper evaluates a number of different aspects of environmental performance at both the firm and the plant level, and discusses the factors which motivate firms to improve their environmental performance as well as the major obstacles which they face.

Using a similar approach, Lilia Domínguez-Villalobos discusses the case of the synthetic fibres industry. This is the only branch of the chemical industry discussed in this volume. Chemical and petrochemical production presents major environmental risks and the leading firms in the industry in developed countries have responded to this through initiatives such as the Responsible Care programme. Given the extensive involvement of TNCs in the chemical industry in Latin America, it is of particular interest as a sector where there is considerable scope for globalisation to lead to improvements in environmental standards.

The final case study by Alfonso Mercado looks at the Mexican iron and steel industry. It provides a point of comparison with the Brazilian steel industry discussed in Chapter 5. Unlike Brazil, relatively cleaner technologies, specifically direct reduction of iron (DRI) and electric arc furnaces (EAFs) as opposed to the integrated route using coke, account for a much higher proportion of steel production in Mexico. Although these differences have important environmental implications, they are not primarily motivated by environmental considerations but rather by the historical trajectory of the industry in the two countries.

As in the other two Mexican case studies, Mercado evaluates the environmental performance of the firms which were interviewed. He contrasts their static performance and their dynamic performance in terms of improvement over time. He then compares the performance of different types of firms and analyses the importance of different drivers and obstacles for firms which perform well in environmental terms and those which do badly on his evaluation.

Notes

1 Manufacturing industry accounted for 21 per cent of the gross domestic product of Latin America and the Caribbean in 1997, compared to a total of only 8 per cent for agriculture, forestry and fishing (World Bank, 1999: Table 4.2).
2 The figures for particulates exclude the contribution of soil and vegetation.

Bibliography

CEC (1996) *Assessing Latin American Markets for North American Environmental Goods and Services*. Montreal: North American Commission for Environmental Cooperation.

Chudnovsky, D., F. Porta, A. Lopez and M. Chidiak (1996) *Los Límites de la Apertura: Liberalización, reestructuración productiva y medio ambiente*. Buenos Aires: CENIT/Alianza Editorial.

Dasgupta, S., H. Hettige and D. Wheeler (1997) *What Improves Environmental Performance? Evidence from Mexican Industry*. Washington, DC: World Bank, Development Research Group, Working Paper Series #1877.

Dasgupta, S., R. Lucas and D. Wheeler (1998) *Small Plants, Pollution and Poverty: New Evidence from Brazil and Mexico*. Washington, DC: World Bank, DECRG Infrastructure/Environment Group (mimeo).

INE (1997) *Programa de Administracíon de la Calidad del Aire del Area Metropolitana de Monterrey, 1997–2000*. Mexico City: Instituto Nacional de Ecología.

INEGI (1995) *Estadísticas del Medio Ambiente: México 1994*. Aguascalientes: Instituto Nacional de Estadística, Geografía y Informática.

OECD (1999) *Environmental Data Compendium, 1999*. Paris: Organisation for Economic Cooperation and Development.

Poder Ejecutivo Federal (1996) *Programa de Medio Ambiente 1995–2000*. Mexico City: Instituto Nacional de Ecología.

Schaper, M. (1999) *Impactos ambientales de los cambios en la estructura exportadora en nueve paises de America Latina y el Caribe: 1980–1995*. Santiago de Chile: CEPAL, División de Medio Ambiente y Asentamientos Humanos.

Shaman, D. (1996) *Brazil's Pollution Regulatory Structure and Background*. Washington, DC: World Bank.

USAID (1995) *Mexico's Environment Markets*. Washington, DC: Center for Environment

World Bank (1995) *Argentina Managing Environmental Pollution: Issues and Options*. Washington, DC: World Bank, Environment and Urban Development Division.

World Bank (1996) *Brazil: Managing Environmental Pollution in the State of Rio de Janeiro*. Vol.1 – *Policy Report*. Washington, DC: World Bank.

World Bank (1999) *World Development Indicators*. Washington, DC: World Bank.

2

GLOBALISATION, TRADE LIBERALISATION AND INDUSTRIAL POLLUTION IN LATIN AMERICA

Rhys Jenkins

Introduction

In recent years two major trends have come into conflict in the international arena. The first is the globalisation of economic activity, defined as 'a *process* in which the structures of *economic markets, technologies,* and *communication patterns* become progressively more international over time' (OECD, 1997: 7). The second is the increased public concern over the environmental impact of economic activity and awareness of the global dimensions of many environmental problems. Whereas globalisation has been associated with increased liberalisation of international markets for goods and capital, growing awareness of environmental problems has led to new forms of regulation both at the national and the international level.

The extent to which globalisation and liberalisation pose a threat to the environment is a matter of heated debate. Advocates of trade liberalisation fear that the environment provides new ammunition for protectionists and will be used to restrain the advance of globalisation. On the other hand, environmentalists see globalisation posing a threat to national environmental standards and believe that the World Trade Organisation (WTO) will prioritise free trade to the detriment of the environment. These conflicting views manifested themselves in the debate over the signing of NAFTA in the early 1990s, and more recently in the criticisms of the WTO which culminated in the protests at the Seattle meeting.

This chapter explores some of these debates empirically by looking at the impact of trade liberalisation on industrial pollution in the three major Latin American countries, Argentina, Brazil and Mexico.[1] Latin America provides a good arena to look at these issues because of the major shift in economic policy regime that took place in the region from the second half of the 1980s.

This is described in the next section. The third part of the chapter provides a conceptual framework for analysing the various ways in which the changed insertion of the Latin American countries into the world economy has affected industrial pollution. The remainder of the chapter considers each of these potential impacts in turn. The picture that emerges is much more complex than either of the extreme positions in the debate would suggest.

Liberalisation in Latin America

The period since the mid-1980s has seen a major liberalisation of the Latin American economies. Although this has included a number of other elements such as privatisation, financial deregulation, tax reforms and changes in the labour market, the most rapid and striking changes have been in the opening of the region's economies through trade liberalisation and changes in policies towards foreign direct investment.

Of the three major Latin American countries, Mexico was the first to open up its economy in the mid-1980s. Argentina and Brazil followed in a wave of countries which began to liberalise around 1990.[2] Trade liberalisation involved the reduction of the average level of import duties, a more uniform level of tariffs (reduced dispersion), a reduction in the proportion of trade covered by non-tariff barriers, and reduced taxes on exports. The average tariff was reduced between the mid-1980s and the mid-1990s from over 50 to 14 per cent in Argentina, from over 80 to 13 per cent in Brazil and from over 40 to 14 per cent in Mexico (Burki and Perry, 1997: Table 2.2; IADB, 1997: Fig.17). By the early 1990s the proportion of items covered by non-tariff barriers was less than 4 per cent in Mexico, 1.5 per cent in Brazil and 0.2 per cent in Argentina (Burki and Perry, 1997: Table 2.2).

At the same time that trade barriers were being dismantled, these economies also relaxed their controls on inward investment. Again Mexico was the first of the three countries to open up to foreign capital. From 1984 the regulatory framework became less restrictive and in 1989 a new set of rules repealed all previous regulations governing foreign investment and widened the range of operations where 100 per cent foreign ownership was permitted. In 1993 a new law was passed which consolidated these changes (Ros *et al.*, 1996). In 1994 the creation of the NAFTA further opened up the Mexican economy to both trade and investment from the USA and Canada.

In Argentina, from 1989 onwards, restrictions on foreign investment in a number of sectors including information technology, telecommunications and electronics were removed. In 1993 a new Foreign Investment Law was approved which further opened up the economy to foreign capital and removed all restrictions on profit remittances (Chudnovsky and Lopez, 1997). By 1994 Argentina had totally deregulated foreign investment (Edwards, 1995: Table 7–10).

In Brazil change was less rapid than in Mexico or Argentina. The new constitution of 1988 in fact imposed more controls on the activities of foreign firms. However, in the 1990s controls on outflows of capital were removed and the entry of foreign firms into the information technology industry was permitted. The reforms to the constitution in 1993 and subsequent amendments approved after 1995 further liberalised policy towards foreign investment (Chudnovsky and Lopez, 1997).

The results of the trade opening were clearly visible in all three countries during the 1990s. The share of trade as a percentage of GDP increased first of all in Mexico after the mid-1980s and then in Argentina and Brazil in the 1990s (see Table 2.1). Similarly, foreign investment as a share of gross fixed capital formation increased substantially in the post-reform period.

Framework for analysis

In analysing the impact of liberalisation on pollution, it is useful to decompose emissions into three separate components (cf. Grossman and Krueger, 1992; Birdsall and Wheeler, 1992). The total industrial emissions of any pollutant (E_j) can be derived as follows:

$$E_j = \sum e_{ij} w_i Y$$

where e_{ij} are the emissions of pollutant j per unit of value added in industry i, w_i is the share of value added of industry i in total industrial value added, and Y is the total industrial value added.

If liberalisation affects pollution then it must be reflected in changes in one or more of the following three components. First, if liberalisation affects the level of industrial activity (Y), then the level of emissions will change. This is what has been termed the *scale* effect. Second, the level of emissions will

Table 2.1 Trade[a] and FDI stock as a percentage of GDP

	1980	*1990*	*1995*	*1997/8*[b]
Argentina – trade	21.5	14.2	22.2	27.8
FDI stock	6.9	5.3	9.9	12.3
Brazil – trade	15.4	14.4	23.2	25.9
FDI stock	7.4	7.8	14.4	15.9
Mexico – trade	30.9	37.9	57.4	76.1
FDI stock	4.2	9.2	14.3	12.5

Sources: trade – Inter American Development Bank; FDI stock – UNCTAD (1999) *World Investment Report 1999*. Geneva: United Nations, Annex Table B6.

Notes

a Total trade (imports + exports) as a share of GDP.

b 1997 for FDI and 1998 for trade.

depend on the contribution of different industries to total value added (w_i). Clearly, where more polluting industries such as petrochemicals or cement are increasing their share of production, total pollution will tend to rise. This is referred to as the *composition* effect. Finally, pollution will change with any reductions in emissions per unit of output which are achieved within an industry (e_{ij}). This has been described as the *process* effect or *technique* effect which comes about from changes in the pollution intensity of each industry.

The impact of liberalisation on each of these three elements cannot be derived a priori. The scale effect depends on whether, in a particular country, liberalisation leads to faster industrial growth. Changes in composition can either increase or decrease the overall level of emissions, depending on the relative growth of different industries, so that the impact of this effect on pollution is ambiguous. Changes in pollution intensity within industries will depend on whether or not firms introduce less polluting technologies or improve their environmental management in response to liberalisation. It may also be affected indirectly as a result of the impact of liberalisation on environmental regulation within the country concerned.

Thus, the outcome in terms of industrial pollution will depend on the direction and relative importance of the impact of liberalisation on the scale of industrial production; on the structure of industrial production; on production technology and management; and on the regulatory framework. Each of these will be discussed in the following sections. One important limitation of this approach is that it concentrates solely on industrial emissions and therefore does not capture the effects of increasing emissions from transporting products on a much greater scale, which is undoubtedly one of the effects of liberalisation.

Scale effects of liberalisation

The *scale effects* of liberalisation on industrial pollution will depend on whether or not it leads to a faster growth of production than would have been the case in the absence of reform. A major rationale for trade reform is the belief that it will lead to a faster rate of economic growth. However, the impact of liberalisation on growth is widely debated and it is by no means universally agreed that greater openness does in fact lead to better economic performance.

Recently, several studies have attempted to estimate empirically the impact of the recent policy reforms in Latin America on economic growth in the region (Edwards, 1995: Ch.5; IADB, 1997; Burki and Perry, 1997; Stallings and Peres, 2000). They all concur that liberalisation has increased growth rates in the region in the 1990s compared to the 1980s. Although the estimates arrived at differ, and they use different methodologies, they generally agree that there has been a perceptible impact on growth rates and that the opening up of the economies has been the most important reform in terms of improving economic performance. This no doubt reflects the fact that it is in the area of trade and investment that reforms were embarked on first of all, and where they have gone furthest.

However, accepting that liberalisation has had positive effects on GDP growth does not necessarily imply that it has had the same effect on the growth of the manufacturing sector which is of primary interest here. First, as can be seen in Table 2.2, manufacturing growth has lagged behind GDP growth in both Argentina and Brazil during the 1990s. Thus, assuming that trade liberalisation has raised the rate of growth of GDP by one percentage point in the 1990s,[3] this would be reflected in an additional growth in manufacturing value added of around three-fifths of a percentage point in Argentina and of three-quarters of a percent in Brazil.

The point though is not just that GDP growth exaggerates the effect of liberalisation on manufacturing in Argentina and Brazil, but that the lag in manufacturing growth may be a *result* of liberalisation.[4] If this is indeed the case, then the effect of liberalisation could be to slow down the rate of growth of manufacturing. This is the argument that liberalisation leads to *de-industrialisation*. Since liberalisation implies a change in the allocation of resources between different economic activities, it is quite possible that when economies are opened up to greater international competition, resources move away from the manufacturing sector which was protected under the previous economic model.

Unfortunately, the studies referred to earlier do not examine the impact of liberalisation on the growth of manufacturing in Latin America so that it is necessary to look at other studies in order to address this question. One study which attempted to do this for several countries in the region concluded that the impact of policy reform on the growth of the manufacturing sector was either negative or not significant (Weeks, 1996). Of the three countries discussed here, Mexico showed a lower rate of growth of manufacturing in the post-reform period, and in Argentina and Brazil there was no clear effect.[5] A

Table 2.2 Elasticity[a] of growth of manufacturing value added (MVA) with respect to GDP

	1970–80	*1980–90*	*1990–97*
Argentina	0.52	2.00[b]	0.59
Brazil	1.11	0.59	0.74
Mexico	1.11	2.14	1.32

Sources: World Bank, *World Development Indicators*, 1999: Table 4.1; World Bank, *World Development Report, 1995*: Annex Table 2.

Notes

a Elasticity is calculated as the ratio of the growth of MVA to the rate of growth of GDP in each period.

b In this period the growth of both MVA and GDP in Argentina was negative.

survey of a number of studies on the impact of trade liberalisation on industrial development in Latin America also concluded that there was no positive long-term effect of liberalisation on either manufacturing productivity or growth (Dijkstra, 1997).

It is difficult to say then what has been the net effect of liberalisation on the growth of the manufacturing sector in the three countries. Since the impact of liberalisation on overall growth is relatively limited, and manufacturing has grown more slowly than the economy as a whole in Argentina[6] and Brazil, the *scale effect* of liberalisation in the two countries is not likely to have significantly increased industrial pollution. Even in Mexico, where manufacturing has grown faster than GDP in the 1990s, a detailed study of manufacturing investment concluded that the effect of the economic reforms of the 1980s had not led to faster growth of the manufacturing sector (Moreno-Brid, 1999). The NAFTA agreement may have had a more significant effect on industrial growth in Mexico in recent years but there is still no detailed evidence on the precise impact which it has had on the growth of the manufacturing sector.

Although liberalisation may not have led to significant increases in the overall level of industrial pollution in the three countries (with the possible exception of the impact of NAFTA on Mexico), there are a number of aspects not considered here which may indeed have caused significant environmental problems. First, liberalisation may lead to changes in the spatial distribution of industrial activity, which can have significant local environmental consequences.[7] The impact of the level of industrial emissions in a country will depend very much on how these are distributed geographically. In Mexico, for instance, it has been estimated that the Federal District and three states account for more than half of all toxic emissions (Mercado and Fernandez, 2000). In the context of liberalisation, even if the overall growth of manufacturing is not much affected, its regional distribution may well be. Thus, in Mexico, industrial growth has been particularly rapid in the *maquila* industry of the USA–Mexico border, giving rise to significant environmental problems in the border cities.

Another aspect, which it has not been possible to consider here, is the impact of the growth of freight transport as a result of liberalisation. There is clearly an environmental effect here where increased trade volumes occur and therefore goods are moved over longer distances. However, since the focus is on industrial emissions, it has not been possible to estimate the extent of additional pollution from transport. It is also important to bear in mind that no attempt is made here to look at the overall environmental impacts of liberalisation beyond manufacturing. Clearly, if sectors other than manufacturing have grown particularly fast as a result of liberalisation, the environmental degradation caused in these sectors would also need to be taken into account in any overall evaluation of the environmental effects of economic reform.

Changes in the structure of production

The second way in which liberalisation can affect the level of industrial pollution in a country is through changes in the structure of production. In other words, greater openness to foreign trade or investment can lead to a more or less pollution-intensive composition of manufacturing production.

Theoretical perspectives

There is considerable controversy over whether increased trade leads developing countries to specialise in 'dirty' industries or whether more open economies have less pollution-intensive industrial structures.[8] One view is that since developing countries have less stringent environmental regulation than higher income countries, and that therefore producers internalise less of the environmental costs of production, they will enjoy a comparative advantage in more polluting industries. Consequently trade liberalisation will tend to have a negative impact on their domestic environment (Copeland and Taylor, 1994).

Against this, however, it has been argued that generally environmental control costs in manufacturing industry are low and that factors other than environmental considerations are more important determinants of comparative advantage (Dean, 1992). In this case it is quite possible that a developing country with a less stringent environmental control system may nevertheless have a comparative advantage in less polluting industries. Where there is a correlation between capital intensity and pollution intensity, countries with a comparative advantage in labour-intensive industries will benefit environmentally from specialising according to their comparative advantage. Indeed pollution will tend to increase in developed countries, because of their specialisation in capital-intensive industries, and be reduced in developing countries (Antweiler *et al.*, 1998). This is associated with the view that the structure of protection in developing countries has a 'brown bias'. In other words, it is suggested that, under import substitution regimes, highly polluting industries tend to receive higher protection than less polluting industries (Birdsall and Wheeler, 1992).

A similar debate exists over the impact of liberalisation of foreign investment on the environment. This has usually been discussed in terms of whether or not developing countries are 'pollution havens' in which lax environmental regulation is used to attract polluting firms, and the corresponding 'industrial flight' of polluters faced with higher standards and/or stricter enforcement in developed countries. A major issue here is the extent to which environmental regulation is considered to be an important determinant of location decisions by foreign investors. Critics of the 'pollution haven' hypothesis claim that other factors such as market opportunities, labour cost and the 'climate for investment' are far more important determinants of the location of production. On the other hand, case study and anecdotal evidence reveal examples of investments where environmental considerations have played an important role (Mabey and McNally, 1999).

Openness, trade liberalisation and pollution intensity in Latin America

Two kinds of previous studies can throw some light on the links between trade policies and industrial pollution in Latin America. One type of study focuses on the pollution intensity of industrial production as a whole and how it varies between countries and over time. The second type of study looks more explicitly at trade and particularly at exports to see how polluting they are, and how this has changed over time.

Within the first group, an influential study carried out in the early 1990s for the World Bank found that more open economies in the region tended to have a lower rate of growth of pollution than more protected economies, leading the authors to conclude that,

> 'pollution havens' can be found, but not where they have generally been sought. They are in protectionist economies.
>
> (Birdsall and Wheeler, 1992: 167).

A critique of this study, however, has pointed to a number of weaknesses (Rock, 1996). First, it only considers toxic emissions which are not necessarily correlated with conventional air or water pollutants. Second, it bases its measure of trade openness on the Dollar Index which has itself been subject to criticism (Rodrik, 1994). Third, Birdsall and Wheeler compare the rate of growth of pollution intensity rather than the absolute level in open and closed economies. Finally, it has been suggested that the lower growth of pollution in more open economies may be a statistical artefact.[9] Rock's own estimates (not just for Latin America) show that more open economies tend to have a more polluting composition of output.

A more recent World Bank study (Mani and Wheeler, 1999) has shown a steady increase in the share of polluting industries in production in Latin America since the early 1960s, which is in marked contrast to the decrease in the share of such industries in Europe, North America and Japan. This implies that the dominant trend in Latin America has been for production to become more pollution intensive. This is supported by a study of Mexico, which found a significant increase in the toxic intensity of production up to the late 1980s (Ten Kate, 1993).

From the point of view of analysing the impact of trade liberalisation on pollution, a major limitation of these studies is that they generally cover a period up to the mid- or late 1980s, or at the very latest the beginning of the 1990s. Since, as was seen above, liberalisation only began in Mexico in the mid-1980s and in Argentina and Brazil at the beginning of the 1990s, it is necessary to look at more recent data in order to evaluate what has occurred in terms of pollution intensity during the 1990s. A second limitation is that all the studies have concentrated on toxic emissions and as was pointed out above, these are not necessarily representative of all forms of pollution. It is therefore of interest to construct indices of pollution intensity for a variety of different pollutants.

A limited number of studies have looked directly at the links between trade and pollution in Latin America, but they have tended to be quite superficial. A World Resources Institute study estimated the total pollution attributable to export production in various Latin American countries, mainly in the early 1990s, concluding tentatively that there is a slight tendency for export expansion to be concentrated in low pollution-intensity sectors relative to high-pollution sectors (Runge *et al.*, 1997). The data used for both Argentina and Brazil were extremely limited; however, in the case of Mexico the share of exports in production of high pollution-intensity industries fell significantly in the early 1990s.

A more detailed recent study of nine Latin American countries by the Economic Commission for Latin America and the Caribbean found that, in absolute terms, there has been a tendency for the volume of exports from 'dirty' industries to increase since the early 1980s in all the countries studied. However, the share of these industries in exports has declined significantly in the late 1980s or early 1990s for all countries apart from Brazil (Schaper, 1999). The composition effect of recent changes in exports has therefore been towards less pollution-intensive industries.

There have also been a limited number of studies of individual countries that have looked at the relationship between export structure and pollution. In Argentina the share of pollution-intensive industries in exports of manufactures both to the OECD countries and to all destinations declined between 1990 and 1994 (Chudnovsky *et al.*, 1996: Table V.1). In Brazil, which is characterised by a relatively high share of 'dirty' industries in its exports, the share increased significantly during the 1980s (UNCTAD, 1999: Table 3). In the case of Mexico, the concerns about the environmental impacts of NAFTA led to a number of studies during the early 1990s which looked at the structure of Mexican trade (Grossman and Krueger, 1992; Low, 1992). These found that Mexican exports were not heavily concentrated in pollution-intensive industries and that US imports from Mexico were not related to pollution abatement costs in the USA.

The picture that emerges from previous studies of trade–environment linkages in Latin America is therefore very mixed. While this is partly because of differences in the methodologies used in the studies, different definitions of 'dirty' industries and different periods covered, it may also reflect differences in the underlying reality of the various countries. In other words there is not necessarily a single model of the impact of trade on pollution which applies throughout the region. In what follows a common methodology is used to analyse the impact of changing patterns of trade in manufactures on industrial pollution in Argentina, Brazil and Mexico.

One limitation of the previous studies on trade and the environment in Latin America is that they concentrate exclusively on exports. However, trade liberalisation has an impact not only on exports but also on imports, since greater openness implies an increase in *both* exports and imports. A further limitation of most of these studies is that they look at changes in the share of a group of industries defined as pollution intensive in total exports of manufactures. A different approach is adopted here in order to be able to estimate the net effects of changes in both exports and imports on the level of pollution within a country.

In order to do this, it is useful to think of emissions as an input into the production process. Although emissions are often regarded as an output of production, they reflect the fact that environmental resources have been used up in production (Rauscher, 1997: 30). Thus, it is possible to think of traded goods as embodying a certain quantity of environmental resources. Lee and Roland-Holst (1993), who used the concept of 'embodied effluent trade', adopt a similar approach. The implication of this approach is that trade flows imply international transfers of environmental resources. If a country's exports generate more pollution than its imports, then that country is a net exporter of environmental resources. Conversely, if it specialises in relatively 'clean' products, then it will be an importer of environmental resources.

A full evaluation of whether or not Argentina, Brazil and Mexico are net exporters or importers of environmental resources would require an analysis of the overall pattern of trade of the three countries, but this would go well beyond the focus of this book on the environmental impact of industry. Here, therefore, attention will be concentrated on the impact of trade in manufactures and whether or not the pollution generated by manufacturing exports is greater or less than that caused by manufactured imports.

In order to do this, estimates were made of the total emissions associated with exports and imports by using IPPS coefficients.[10] These coefficients were applied to figures for imports and exports of manufactures, reclassified by the four-digit level of the International Standard Industrial Classification, which were made available by the Economic Commission for Latin America and the Caribbean (ECLAC). This generated estimates of the total volume of each pollutant generated by exports, and the pollution avoided (or rather generated elsewhere) as a result of imports. The best way of seeing whether a country is specialising in relatively polluting industries or not, and thus whether it is likely to be a net exporter of environmental resources, is by comparing the average pollution intensity of a dollar of exports with that of a dollar of imports. Table 2.3 shows the ratio of pollution per dollar of exports to imports for each pollutant at the beginning of each country's liberalisation process.

Table 2.3 Relative pollution intensity of exports to imports at the start of liberalisation

	Argentina (1990)	*Brazil (1990)*	*Mexico (1985)*
Toxics	0.54	0.70	0.88
Metals	1.21	2.17	1.31
SO_2	1.47	1.77	1.78
NO_2	1.02	1.29	1.42
CO	1.03	2.03	1.16
VOC	0.90	0.97	1.43
PM10	4.69	4.50	3.48
TSP	3.19	2.56	1.80
BOD	0.70	1.08	0.65
TSS	1.66	4.39	0.74

Abbreviations: VOC, volatile organic compounds; TSP, total suspended particulates; TSS, total suspended solids.
Source: own elaboration from ECLAC trade data and IPPS coefficients.

Prior to liberalisation in each country, exports were more pollution intensive than imports for most of the pollutants covered. The exceptions were toxic emissions where imports were more toxic than exports for all three countries, volatile organic compounds in Argentina and Brazil, biochemical oxygen demand in Argentina and Mexico, and total suspended solids in Mexico.

The *composition effect* of trade liberalisation can be seen by comparing the pollution generated by exports and imports at the beginning of the reforms with that found in 1996, the latest year for which trade data were available. In other words, this effect is reflected in changes in the net transfer of environmental resources between the Latin American countries and the rest of the world. The question is then whether or not liberalisation has led to a greater or lesser transfer of environmental resources from the three countries.

A complicating factor is that changes also occur in the balance of trade from year to year and these will affect the actual transfer of environmental resources in any particular year. In order to focus on the effects of changes in the composition of exports and imports, it is necessary to abstract from changes in the trade balance. Imports of environmental resources were therefore calculated as though trade in manufactures were in balance in both the initial and final year.[11]

Table 2.4 shows whether or not the net transfer of environmental resources, as indicated by each of the pollutants, increased or fell between the start of liberalisation and 1996.

The contrast between Argentina and Brazil on the one hand, and Mexico on the other, could not be starker. In the two South American countries, nine of the ten pollutants showed an increase in the net outflow of environmental

Table 2.4 Impact of trade changes on domestic pollution after liberalisation

	Argentina (1990–6)	*Brazil (1990–6)*	*Mexico (1985–96)*
Toxics	+	+	– –[a]
Metals	+	+	–
SO_2	+	+	–
NO_2	+	+	– –[a]
CO	+	+	–
VOC	+	–	–
PM10	+ +[a]	+	+
TSP	+	+	–
BOD	+	+	– –[a]
TSS	–	+	–

Source: as for Table 2.3.
Note
a + + and – – indicate substantial changes relative to estimates of total industrial pollution at the beginning of the period.

resources[12] following liberalisation, whereas in Mexico, the same number showed a reduction in transfers, the exception being emissions of fine particulates. Thus, while Mexico seems to support the view that trade liberalisation leads to specialisation in relatively less polluting industries, Argentina and Brazil appear to have shifted resources into more polluting activities after opening up their economies.

To put the changes into perspective, they were compared to estimates for the total emissions of the different pollutants at the beginning of the period of liberalisation.[13] Where the change in emissions associated with changing trade patterns was more than 10 per cent of the estimated emissions prior to liberalisation, this was regarded as substantial. This shows that, with the exception of PM10 in Argentina, the increase in emissions in the two South American countries over the period was relatively small. In Mexico, admittedly over a somewhat longer period, three pollutants showed reductions of more than 10 per cent compared to the initial year.[14]

As indicated above, one of the reasons why some authors predict that trade liberalisation would lead to less polluting production in developing countries is because they believe that under import substitution, protectionist policies tended to favour 'dirty industries' and that this 'brown bias' would be removed. How far was this true of the three Latin American countries in the period before the recent trade opening?

In order to answer this question, the structure of protection in Argentina, Brazil and Mexico prior to the period of liberalisation was analysed. Appropriate sectoral estimates of the effective rate of protection (ERP) were obtained for 1973 for Brazil (Coes, 1991: Table 4.1), 1980 for Argentina (Cavallo and Cottani, 1990: Table 3.19) and 1979 for Mexico (Ten Kate and de Mateo Venturini, 1989: Table 4). On the basis of these data, industries were classified into those with high and low levels of effective protection.[15] In Brazil and Mexico industries were considered to have high levels of protection if the ERP was over 50 per cent, while in Argentina where the overall level of protection was higher, an industry was classified as having high ERP when it was over 75 per cent.

Having classified the two groups of industries, it was then possible to calculate the average pollution intensity for high and low ERP industries, and then to derive a ratio between the pollution intensity of the two groups. If protection has a 'brown bias', i.e. tends to protect the more highly polluting industries, then the ratio would be greater than one. Surprisingly, in the case of Argentina and Brazil, the bias was in the opposite direction with the most heavily protected industries having relatively low emissions of most pollutants (see Table 2.5). Only Mexico conformed to expectations, with highly protected industries being relatively pollution intensive with the exception of some total suspended solids and particulates.

This helps to explain why, of the three countries, only Mexico conformed to the pattern of increased specialisation in less polluting industries after liberalisation got under way. This simply reflects the fact that it was the only one of the three countries in which protection had a strong bias in favour of more polluting industries prior to liberalisation.

Foreign investment and pollution intensity

The literature on the impact of foreign investment on pollution intensity in Latin America is even more limited than that on trade, and the evidence is largely anecdotal or based on selective case studies. There are examples where environmental regulation has been an important factor leading companies to relocate to Latin America. A much quoted instance is the relocation of

Table 2.5 Ratio of pollution intensity in high versus low ERP industries

	Argentina	*Brazil*	*Mexico*
Total toxics	0.32	0.17	1.42
Total metals	1.11	0.05	1.60
BOD	0.81	1.32	1.34
TSS	0.30	0.01	0.41
NO_2	0.13	0.39	2.30
PM10	0.08	0.16	0.18
SO_2	0.24	0.19	3.64
CO	0.23	0.11	2.01
TSP	0.21	0.44	0.98
VOC	0.15	0.63	3.27

Abbreviations: TSP, total suspended particulates.
Source: own elaboration from IPPS and UNIDO data.

furniture manufacturers from the Los Angeles region to Mexico in the late 1980s (OTA, 1992: 100).

However, more aggregate studies have not generally found a clear pattern. A study of Mexico and Venezuela found no correlation between the sectors in which foreign investment was located and either pollution abatement costs or various indicators of pollution intensity (Eskeland and Harrison, 1997). Another area of debate has been in relation to the *maquiladoras* in Mexico where contrasting views on the importance of environmental costs as a factor influencing industrial location have been presented by Grossman and Krueger (1992) who find no link between the pattern of investment and pollution abatement costs, and Molina (1993) who criticises this finding.

If indeed relocation of industry from countries with stricter environmental regulation is an important factor, then it might be expected that foreign firms tend to concentrate in relatively pollution-intensive industries. Argentina, Brazil and Mexico have all received substantial inflows of direct foreign investment in recent years. Unfortunately, it was not possible to obtain recent data on the distribution of foreign ownership by industry, except for Mexico. However, earlier data can be used to give an indication of the kinds of sectors in which foreign capital has tended to be concentrated (see Table 2.6).

In order to test whether or not foreign firms tend to concentrate in the more pollution-intensive industries in the three countries, Spearman's rank correlations were calculated between the share of foreign ownership in an industry and emissions per dollar of value added for the pollutants listed in Table 2.5 above. None of the rank correlations calculated in this way were significant at the 5 per cent level in any of the three countries. Thus, there is no evidence to support the view that foreign capital has tended to concentrate in 'dirty' industries.

Although this evidence does not indicate whether or not there has been a shift in the composition of foreign investment towards more or less polluting industries in the period of the economic reforms, it does suggest that factors other than environmental considerations are the major determinants of the sectoral distribution of foreign capital. It seems unlikely therefore that liberalisation has had significant beneficial or negative effects on pollution in terms of the sectoral distribution of foreign investment.

Conclusion

The evidence shows that there is no universal pattern as far as the composition effect of liberalisation on pollution in the three countries is concerned. Only in Mexico, where the previous pattern of protection was biased in favour of relatively polluting industries, has greater openness been associated with a shift in comparative advantage towards cleaner industries. In both Argentina and Brazil, where polluting industries were not more protected than others, this has not been the case. Moreover, in none of the three

Table 2.6 Share of foreign firms in manufacturing in Latin America

ISIC	*Industry*	*Argentina (1983) (%)*	*Brazil (1983) (%)*	*Mexico (1994) (%)*
311	Food products	21.2	18.0	15.2
313	Beverages	63.3	15.0	15.0
314	Tobacco	99.6	73.0	42.0
321	Textiles	22.4	22.0	21.9
322	Wearing apparel, except footwear	6.3	4.0	7.6
323	Leather products	16.0	15.0	7.0
324	Footwear, except rubber or plastic	47.8	4.0	5.0
331	Wood products, except furniture	16.5	5.0	5.8
332	Furniture, except metal	32.9	3.0	7.3
341	Paper and products	66.9	21.0	27.0
342	Printing and publishing	8.3	3.0	8.8
351	Industrial chemicals	83.4	21.0	21.9
352	Other chemicals	57.4	62.0	60.5
353	Petroleum refineries	39.9	N/A	0.0
354	Misc. petroleum and coal products	80.1	N/A	25.0
355	Rubber products	69.2	63.0	31.6
356	Plastic products	58.8	17.0	23.7
361	Pottery, china, earthenware	6.2	N/A	17.7
362	Glass and products	62.1	N/A	18.9
369	Other non-metallic mineral products	28.6	N/A	4.8
371	Iron and steel	28.2	23.0	18.9
372	Non-ferrous metals	34.8	44.0	31.4
381	Fabricated metal products	22.1	23.0	14.1
382	Machinery, except electrical	74.3	41.0	52.5
383	Machinery (electric)	74.0	44.0	58.3
384	Transport equipment	88.4	68.0	64.4
385	Professional and scientific equipment	98.5	N/A	64.3
390	Other manufactured products	N/A	29.0	28.5
	Total	N/A	28.5	28.5

Sources: Argentina – Basualdo *et al.* (1988): Table 18; Brazil – Fritsch and Franco (1991): Table 1.6; Mexico – INEGI.

countries does there seem to have been any link between the pattern of foreign investment in the manufacturing sector and pollution intensity.

Liberalisation and production processes

Theoretical perspectives

Perhaps more important than changes in the allocation of resources between different industries, from the point of view of the impact of liberalisation on pollution levels, are the changes which take place within industries at the levels of firms and plants. A number of arguments have been put forward here concerning the potential links between liberalisation and environmental performance.

It has been claimed that producing for export leads to the adoption of clean technologies because of the requirements of international markets. Cases in point arise for instance where chlorine needs to be eliminated in the pulp and paper industry, and chromium in tanning. Even where there are no such specific requirements, producers may fear that protectionist lobbies in their main markets will cite low environmental standards as grounds for action against their exports. Thus, exporters may become particularly interested in obtaining ISO 14000 certification in order to demonstrate their environmental credentials. It is also argued that more open economies have better access to the latest foreign environmental technology and that liberalisation will, therefore, lead to more rapid diffusion of less polluting production methods.

Despite these potential gains from producing for export and having access to imported equipment, increased competition which forces cost minimisation, whether to compete in export markets, or against imports in the domestic market, may make it more difficult to adopt measures to protect the environment. The question of competitiveness is at the heart of many of the environmental concerns surrounding trade liberalisation. The more subject an industry is to international competitive pressures, the more resistant it is likely to be to attempts by regulators to impose environmental protection measures which will increase costs.

Arguments concerning the beneficial effects of liberalisation have also been put forward in the case of foreign investment. It is often claimed that TNCs adhere to their own corporate environmental standards which are higher than those of the developing countries in which they operate (Gladwin, 1987). Thus, increased inflows of foreign capital tend to bring with them higher environmental standards. The extent to which multinationals do in practice require their subsidiaries to observe higher environmental practices is unclear and two surveys of such firms came to quite different conclusions.[16] Even when multinational companies do not have explicit corporate environmental policies, their tendency to use parent company technology, which has been developed to meet the stricter regulatory requirements of their home countries, will lead to them having less polluting production than local firms in developing countries (Ferruntino, 1995).

Trade liberalisation and cleaner production

The evidence from previous studies on the link between production for export and pollution is not conclusive. Some of it is based on case studies and there are undoubtedly examples of firms reducing emissions in order to meet foreign product standards, e.g. in the pulp and paper industry in Chile (Birdsall and Wheeler, 1992). More systematic studies involving surveys of environmental management by firms also provide evidence on the relative behaviour of firms which produce for export compared to those which produce mainly for the domestic market. Contrary to expectations, a study of 90 firms in the Mexico City Metropolitan Area did not find a statistically significant relationship between the proportion of output exported and the degree to which the firm protected the environment (Dominguez, 1999). This was confirmed by an econometric study of 236 firms in Mexico which found that there was no link between producing for export to OECD countries and the environmental performance of firms (Dasgupta *et al.*, 1997).

In Brazil, in contrast, a study of firms in the chemical industry found a positive association between producing for exports, whether in absolute terms or as a share of output, and participation in the industry's 'responsible care' programme (Roberts, 1998). Case studies of trade and environment issues in Brazil suggest that, whereas production for export has forced firms to meet higher environmental standards in the pulp and paper industry, there have as yet been no such pressures on the iron and steel industry (UNCTAD, 1999).

Case studies of three industries in Argentina also showed a mixed picture. A major producer of pulp had indeed responded directly to demands in the European market by improving its environmental performance, and a packaging firm had also taken steps to meet German requirements on using recycled materials and making it possible to reuse its products (Chudnovsky *et al.*, 1996: Ch.VI). One steel company had also been inspected by a European customer interested in its environmental performance, and the leading firms in the industry took the view that they needed to meet international quality, safety and environmental norms if they were to be competitive in the most demanding markets (Chudnovsky *et al.*, 1996: 502). In contrast, in the petrochemical industry, demands from export markets had not been a major factor leading firms to adopt environmental measures (Chudnovsky *et al.*, 1996: Ch.VII).

The studies in this book provide further evidence on the link between exporting and environmental performance. Chudnovsky *et al.* find that, for a small sample of large firms in Argentina, export-oriented firms have gone much further in terms of environmental management and pollution prevention than those which produce mainly for the domestic market (see Chapter 4).

The case studies of Brazil present rather different pictures, although without a formal comparison of firms according to their major markets. Dalcomuni finds that in the pulp industry the demands of European customers have been a major factor in improving environmental performance in the industry (see Chapter 7). In contrast, there is no clear link between producing for export and environmental performance in the steel industry (see Chapter 5). In the leather industry, while some chemicals have been substituted in order to be able to export to the German market, in general, the lack of a market for 'ecological' leather has meant that exporters have not been particularly concerned about the environmental impacts of their activities. Moreover, the fact that an increasing proportion of export production is of relatively pollution-intensive 'wet blue' means that liberalisation has tended to lead to more polluting leather exports (see Chapter 6).

The Mexican studies also show a mixed picture. In the case of textiles, the firms which are most export-oriented tend also to have the best indicators of corporate environmental performance, although at the plant level the differences between exporters and non-exporters are not significant (see Chapter 8). Paradoxically, in the case of synthetic fibres, there appears to be an inverse relation between the share of output exported and environmental performance, although Dominguez warns that this result should be treated with caution (see Chapter 9). However, it is clear that the demands of clients in the export market have not been a particularly important driver for environmental improvement in either of these two industries. In the Mexican steel industry, firms which produce virtually entirely for the domestic market have the worst environmental performance, although it is not the case that the most export-oriented firms perform better than those with an intermediate level of exports (see Chapter 10).

The evidence suggests that there are important inter-industry differences in the impact of producing for export on environmental performance. In some industries, of which pulp and paper is the most notable example, the demands of the export market lead to exporters adopting cleaner technologies and improving their environmental management. In other industries, however, there is little difference between the environmental performance of exporters and producers for the domestic market. What differences do exist may well have more to do with other factors such as the size of the firm or its technological level rather than exports *per se*. On the other hand, the situation is by no means static and it is quite possible that as environmental concerns grow in developed country markets, and firms begin to give more emphasis to the environmental impacts of their suppliers, exporters will increasingly have to adopt higher environmental standards.

The impact of liberalisation on pollution through better access to imported environmental equipment is difficult to evaluate. This is partly because, although some capital equipment can be classified as being 'environmental', this is by no means clear-cut. There is both a conceptual and a classificatory problem which makes it difficult to identify changes in imports of equipment which have environmental benefits. At the conceptual level, improvements in environmental performance are often embedded in improved processes and cannot be identified with particular pieces of equipment. Even where specific items of equipment can be identified as 'environmental', conventional trade classifications do not necessarily separate these from other 'non-environmental' equipment, and certainly do not aggregate them together to provide an easily accessible measure of import of environmental equipment.

Consequently there are very few data available on imports of environmental equipment to Latin America. Estimates from the early 1990s show that between 19 and 25 per cent of the total environmental market in Argentina, Brazil and Mexico were supplied by imports (Barton, 1998: Table 8). However, the share of the environmental market in GDP in each country has remained relatively small (between 0.6 and 0.7 per cent) compared to the OECD countries (Schaper, 1999: Table IV.2). This implies that imports of environmental equipment account for only a very small share of total imports in each country. There is little evidence from detailed case studies that access to imported equipment has been an important factor in determining the environmental performance of firms in Latin America.

Foreign investment and cleaner production

Empirical evidence from Latin America on the impact of foreign investment on the pollution intensity of production is relatively limited. Despite the a priori arguments suggesting that foreign owned firms may perform better than domestic firms, convincing empirical evidence to show that ownership is a key factor is lacking so far.

Two different types of approaches have been used to look at this issue. The simplest involves comparing the proportion of foreign firms who are regarded as good performers in environmental terms with those who are poor performers. These generally find that foreign subsidiaries are over-represented amongst the best performers. Thus, a study of Mexico found that foreign owned firms accounted for 56 per cent of the best environmental performers, although such firms made up only 26 per cent of all firms in the sample (Dominguez, 1999: Table IV.18). The Argentinian study in this volume (see Chapter 4) also found that foreign owned firms were significantly better represented than domestic firms amongst those which were most advanced both in terms of environmental management and pollution prevention. Similarly, the previously quoted study of the Brazilian chemical industry found that foreign firms were disproportionately represented amongst the firms with 'responsible care' programmes and with ISO 14000 certification (Roberts, 1998).

However, this approach does not separate out the impact of ownership *per se* on environmental performance since it does not control for other variables which may also influence performance such as size of firm, vintage of equipment or principal market. Studies which have attempted to take other factors into account using econometric techniques have failed to find a significant relationship, although a simple comparison of the proportion of foreign owned firms amongst those with the best environmental performance might have suggested that foreign ownership was an important factor. Thus, the previously quoted study of Mexico found that overall the relationship between foreign ownership and environmental performance was not statistically significant (Dominguez, 1999). Similarly, a larger Mexican study also failed to find any relationship between foreign ownership and environmental performance (Dasgupta *et al.*, 1997).

Some of the studies in this volume also provide some evidence on the comparative environmental performance of foreign and local firms, although a number do not have enough foreign subsidiaries to make meaningful comparisons. Because of the small number of firms involved, none of the studies use any kind of multivariate analysis. The results are mixed. In the case of the Mexican synthetic fibres industry, foreign firms tend to have a better environmental performance than domestic firms, while in the steel industry there was no evidence of superior static performance by foreign subsidiaries, although they did perform better on the dynamic indicators. These studies, however, suffer from the same limitation as most previous studies, i.e. that they do not control for other factors, so that the case that foreign ownership is a key factor remains unproven.

Conclusion

This section has considered a number of ways in which liberalisation might have a *process effect* on industrial pollution and has presented some evidence from the major Latin American countries. The picture that emerges is a mixed one. In some industries, the demands of the export market have led to firms improving their environmental performance, but so far this appears to be limited to industries that have been a target for environmental activists and consumer pressure, such as pulp and paper, and the effect is not sufficiently pronounced to be obvious in surveys of a large number of firms in a range of industries. In the future, as environmental concerns grow and are increasingly linked to trade issues, the differences between exporters and firms producing for the domestic market may well widen.

The evidence concerning the impact of liberalisation on imports of cleaner technologies is difficult to evaluate. There is little doubt that imports of capital goods have risen sharply following liberalisation and this potentially could lead to the incorporation of cleaner technologies. However, more direct evidence of this is hard to come by, and environmental investments in Latin America continue to be well below the levels of the advanced industrial countries.

Finally, as was noted above, despite claims concerning the high environmental standards required by multinational corporations, there is no clear evidence that foreign owned firms perform better than domestic ones, once factors such as size and industry are taken into account. Thus, the liberalisation of foreign investment does not necessarily lead to improved environmental performance in host countries.

However, comparisons between the environmental performance of different groups of firms only partially address the question of the impact of greater openness on the environment. One of the arguments concerning the possible negative environmental impact of liberalisation of trade and investment is that it leads to weaker environmental standards. If this were indeed the case then the impact would be felt equally by all firms whether they were exporters or producers for the domestic market, foreign or locally owned. In this case a lack of a clear difference between two groups of firms does not necessarily mean that increased trade or greater openness to foreign investment has not had an important environmental effect. It is thus necessary to look at the impact which liberalisation has had on the regulatory framework in Latin America.

Liberalisation and environmental regulation

One of the major concerns voiced by those who fear the environmental consequences of liberalisation is that increased international competition will lead to a lowering of environmental standards – the so-called 'race-to-the-bottom'. In other words as trade is liberalised, producers in the countries with the lowest environmental standards will enjoy a cost advantage because they have internalised less of the environmental costs of their activities. This will lead firms in countries with higher environmental standards to put pressure on their governments to relax these standards in order to ensure a 'level playing field'. The result of this process will be to put downward pressure on standards generally. A less extreme outcome is that increased international competition keeps environmental policy initiatives 'stuck-in-the-mud'. Thus, rather than reducing environmental standards, it has a 'chilling effect' on attempts to increase environmental protection (Mabey and McNally, 1999).

The empirical evidence on these issues is at present mainly anecdotal.[17] There are certainly examples where companies have lobbied successfully to prevent the introduction of environmental measures such as energy taxes, through appealing to the competitive effects that their introduction would have. Two recent surveys conclude that liberalisation of trade and investment has led to a reduction in regulatory autonomy, or at least a perception that this is the case, which leads to a need for greater international co-operation (Nordström and Vaughan, 1999; Mabey and McNally, 1999).

There is no evidence to support the view that liberalisation has led to a 'race-to-the-bottom' in environmental regulation in Argentina, Brazil and Mexico. In all three countries the trend in recent years has been for increased environmental regulation, rather than for standards to be reduced as trade and investment have been liberalised.

Argentina has been the slowest of the three countries in developing its environmental regulation. Although water standards were set from the late 1940s and these were extended to air quality in the 1970s, it is only in the 1990s that other initiatives have been developed with the creation of the Secretaria de Recursos Naturales y Ambiente Humano in 1991 and the passing of a law on hazardous wastes in 1992. However, in the mid-1990s the World Bank described the situation in the following terms:

> Government institutions charged with environmental policies are weak, their responsibilities are fragmented, and enforcement is inadequate in many areas. The institutional framework for environmental management involves a web of overlapping national, provincial and municipal agencies. The resulting unusually complex system of laws, regulations and authorities has led to unevenness and uncertainty in the enforcement of regulations, and opened many opportunities for polluter to evade compliance with environmental objectives.
>
> (World Bank, 1995: 5)

Although progress is slow in Argentina, given the previous low levels of environmental regulation there is no evidence of standards being reduced. The 'regulatory chill' hypothesis is more difficult to dismiss since the limited improvement in regulation might be a result of fears for the competitive position of firms. However, it is more likely that it reflects the low priority given to environmental issues in Argentina and the relative weakness of environmental pressure groups.

Environmental regulation in Brazil is characterised by considerable differences between different states, although minimum standards are set at the federal level (see Chapters 5 and 6). Environmental issues have had a rather higher profile in Brazil than in Argentina. The first attempts at systematically controlling industrial pollution came in 1974 with the Industrial Pollution Control and Prevention Law. In 1988 the new federal constitution reorganised the country's environmental system and a Secretariat for the Environment (SENAM) was created (Shaman, 1996). In 1992, however, a World Bank report highlighted the limited enforcement capacity of Brazil's environmental agencies, and the poor co-ordination between the federal, state and municipal levels. In the case of FEEMA, the environmental agency for Rio State, the late 1980s and the 1990s saw a drastic decline in its capacity, caused by a lack of political support and limited budgets (World Bank, 1996). However, this seems to reflect specific local circumstances rather than being part of a generalised decline in environmental standards associated with liberalisation.

As in the case of Argentina, there may be a 'regulatory chill' effect of competition on environmental regulation in Brazil. Odegard (Chapter 6) reports that regulators tend to be less demanding of the tanners when they are in economic difficulties. The trend to proceed slowly in terms of regulation and enforcement is accentuated by competition between different states who perceive themselves facing:

> a trade-off between revenue and environmental quality . . . the better a state's enforcement of federal requirements, the less polluted the state is but the lower is the state net revenue. . . . This sort of reasoning may explain why the degree of enforcement of pollution laws is often negotiated between polluter and state. . . .
>
> (Shaman, 1996: 4, quoting a World Bank report)

Mexico is the most interesting of the three cases and offers a somewhat different scenario from that of either Argentina or Brazil. This is because it has not only undertaken unilateral trade liberalisation like the other two large Latin American countries, but has also entered into a regional free trade agreement with the USA and Canada.

The first attempt at environmental policy making in Mexico was the 1971 Federal Law for the Prevention and Control of Environmental Contamination, but this had little effect and in 1982 a new law was promulgated to strengthen environmental legislation and a year later the Secretaria de Desarrollo Urbano y Ecologia (SEDUE) was created. In 1988 this was replaced by the General Law for Ecological Equilibrium and Environmental Protection. Despite the earlier laws, it was only in the 1990s that enforcement of environmental regulation was stepped up in Mexico (Ruijters, 1995: OECD, 1998).

There is little doubt that the increased environmental protection measures taken by Mexico in the early 1990s were partly a response to the concerns being expressed in the USA at the time over the implications of the NAFTA agreement with Mexico (Hogenboom, 1998: Ch.6). The negotiations, which began in 1990, mobilised environmental groups in the USA on an unprecedented scale around the possibility that NAFTA would lead to 'environmental dumping' and 'industrial flight' of firms to Mexico. This led to an environmental side-agreement being signed and to the setting up of the Commission for Environmental Co-operation.

Thus, in the Mexican case, liberalisation has not been associated with a lowering of environmental standards, but quite the opposite, particularly in the field of enforcement. It is important to highlight two specific characteristics of the Mexican experience which help account for this. First, it is not liberalisation *per se* that has led to higher environmental standards but specifically the entry into a regional agreement. Unilateral liberalisation in the late 1980s did not lead to a major shift in environmental regulation, which only began when the NAFTA was under negotiation. Second, it is important to bear in mind that the particular integration scheme involved Mexico joining with countries which already had much higher environmental standards, and which might therefore tend to pull Mexican standards upwards.

Reviewing the evidence for the three countries therefore, there is no reason to believe that liberalisation has led to a lowering of environmental standards in any of them. In the case of Argentina and Brazil, it is possible that an element of 'regulatory chill' may exist, but it is difficult to establish that, in the absence of liberalisation, there would have been a more rapid increase in environmental regulation. In the case of Mexico, however, the evidence is that NAFTA has contributed towards improved environmental regulation.

Conclusion

This chapter has analysed the impact of recent liberalisation of trade and investment in the three main Latin American countries on industrial pollution. A major constraint of any such exercise is the lack of directly observed data on emissions by the industrial sector in any of the countries concerned. Consequently it has been necessary to approach the issue in a more roundabout manner.

The literature has identified a number of potential impacts that liberalisation may have on pollution and each of these was considered in turn. Table 2.7 provides a summary evaluation of the probable effect of liberalisation under each of the headings considered in this chapter.

The *scale effect* of liberalisation is probably quite limited in all three countries. Even if overall liberalisation has led to increased growth as some writers have suggested, it is by no means clear that this has led to a faster rate of growth of the manufacturing sector. Indeed, manufacturing has lagged behind GDP growth in both Argentina and Brazil, so that only in Mexico is it likely that the combined effect of liberalisation and NAFTA have led to a more rapid rate of growth of manufacturing and hence of industrial pollution.

The direction of the *composition effect* of liberalisation differed between countries. While in Argentina and Brazil, changes in exports and imports after

Table 2.7 Impact of liberalisation on industrial pollution

	Argentina	*Brazil*	*Mexico*
Scale	Small – possible slight increase in emissions	Negligible	Negligible from liberalisation but could be significant from NAFTA
Composition	Small increase in emissions	Small increase in emissions	Reduced emissions
Process	Reduced emissions in some sectors	Reduced emissions in some sectors	Reduced emissions in some sectors
Regulation	Possible 'chilling' but effect small	Possible 'chilling' but effect small	Increased regulation with NAFTA
Overall	Slight increase in emissions	Negligible effect	Reduced emissions

liberalisation led to an increased transfer of environmental resources to the rest of the world, in Mexico the opposite occurred. This indicates that there is no general relationship between trade liberalisation and the pollution intensity of trade but that it will depend on the comparative advantage of the country concerned and the pattern of protection prior to liberalisation. However, the overall contribution of changes in trade flows to industrial pollution in the two South American countries after liberalisation was relatively small.

The *process effects* of liberalisation are not clear-cut and differ from industry to industry. Thus, whereas production for export encourages environmental improvements in some industries, the overall impact may be limited. Similarly, the role of foreign firms in bringing in cleaner technologies and better environmental management does not come out in aggregate studies where other factors such as firm size and industry are taken into account. Finally, although there is a potential positive effect as a result of better access to imported environmental equipment, this is likely to have been quite small in practice.

Finally, there is evidence in Mexico that NAFTA has led to a raising of environmental standards. Liberalisation has had no such effect in either Argentina or Brazil and may have contributed to the relatively slow rate of improvement, although by its very nature this is difficult to assess.

Looked at overall it is clear that the extreme positions concerning the effects of liberalisation on the environment have not been vindicated. The optimism of those who see a 'win–win' relationship whereby trade liberalisation produces substantial environmental benefits is not borne out. The outcome is likely to differ from country to country depending on where its particular comparative advantage lies, and the structure of protection before liberalisation. Nor have the fears of those who see liberalisation as a major threat to the environment been vindicated. If such fears are realised it is likely to be through the impact on environmental regulation rather than through massive relocation of 'dirty' industries to Latin America. The Mexican experience shows, however, that increased international integration does not have to lead to a lowering of domestic environmental standards.

The general conclusion to be drawn from the evidence presented here is that the focus of the debate on industrial pollution in Latin America should not be primarily about the implications of different economic policies for the environment but rather about the design of appropriate environmental policies and regulatory frameworks to reduce pollution. Whether countries adopt more outward- or inward-looking policies, they need to have in place a strong regulatory system if industrial pollution is to be controlled. In the absence of such a system, economic growth is likely to lead to further environmental degradation whether it is driven by international or domestic factors.

Notes

1 The focus of this chapter and the rest of this book, with the exception of Chapter 3, is on industrial pollution. However, it should be borne in mind that liberalisation also has much wider environmental consequences than just those associated with manufacturing, as for example where it leads to increased exploitation of natural resources. The chapter should not be read as analysing the overall environmental implications of the recent opening of the Latin American economies.

2 Different authors give different exact dates for the start of trade liberalisation in each country, but they are all agreed that the key dates are around 1985 in Mexico and 1990 in Argentina and Brazil. See Agosin and Ffrench-Davis (1993: Table 1), Edwards (1995: Table 1-1) and Burkis and Perry (1997: Ch.II.A).

3 Unfortunately, the studies do not give estimates of the contribution of trade liberalisation to overall growth at the level of individual countries. For the region as a whole, it is estimated that trade reforms increased growth by 0.8% calculated as a simple average and 1.1% as a weighted average (IADB, 1997: Table 5).

4 This is suggested in the case of Brazil by the much lower elasticity of Manufacturing Value Added (MVA) in the post-reform period (1990–7) compared to the pre-crisis period (1970–80). In the case of Argentina, the elasticity was quite similar in the two periods, but it has to be remembered that the country went through an earlier period of significant trade liberalisation in the second half of the 1970s, which was subsequently reversed, and that this affected the growth of manufacturing in the 1970–80 period.

5 A limitation of this finding is that Weeks only considered the period up to 1992 which did not leave much time for the impact of trade liberalisation to be observed in Argentina and Brazil.

6 The publication of the 1994 Industrial Census in Argentina led to a further re-evaluation of the performance of manufacturing after liberalisation since it showed that the growth of value added since 1990 had been negligible (Kosacoff, 2000: Ch.6).

7 I am indebted to David Barkin for drawing this point to my attention.

8 For a recent review of this debate see Nordström and Vaughan (1999: Ch.III).

9 Specifically it is pointed out that the variable used to measure trade orientation is an interaction term between the rate of growth of per capita income and the so-called Dollar Index of trade orientation. Rock (1996) suggests that what this variable is picking up is the impact of the growth rate rather than trade orientation.

10 These are coefficients which have been calculated by the World Bank on the basis of US data which measure emissions of a wide range of pollutants per US$ of output. While the use of coefficients derived from the USA is likely to underestimate the level of pollution associated with a given level of output in Latin America where environmental regulation is less strict and pollution abatement more limited, nevertheless providing that the ranking of industries in terms of pollution intensity is similar to that in the USA, calculations based on these coefficients are useful in indicating the impact which changes in trade flows and industry structure are likely to have on emissions.

11 This involved applying the average emission coefficient for each pollutant estimated from the composition of imports in the initial and final year, to the total value of exports in that year. This gave an estimate for each year of the net transfer of environmental resources which would have occurred had trade been balanced, and made it possible to calculate the change in net emissions, abstracting from the influence of changes in the trade balance.
12 This implies either a larger surplus in terms of environmental resources exported, or a smaller deficit.
13 The estimates of total emissions were those used to illustrate the growth in industrial pollution in the three countries cited in Chapter 1. They were calculated using data for value added at the three-digit level of the International Standard Industrial Classification (ISIC) and IPPS value added coefficients for emissions and are therefore not strictly comparable to the estimates made from the trade data.
14 These pollutants all showed a decrease of over 10% between 1990 and 1996 as well, so that the longer period was not a key factor in their substantial decline.
15 Some industries were omitted due to the absence of estimates of effective protection. Also, in some cases, estimates were only available at the two-digit level and it was assumed that the three-digit industry shared the same characteristics in terms of protection as the two-digit industry to which it belonged.
16 The majority of TNCs surveyed by the UNCTC Benchmark Corporate Environmental Survey reported having corporate environmental policies which went beyond those required by national legislation of the host country (UNCTC, 1992: 234). However, a Ministry of International Trade and Industry (MITI) survey of Japanese TNCs found quite the opposite, with the majority of firms only taking the measures required to meet local environmental standards (World Bank, 1993: Box 3.2).
17 Only one more formal study exists, by Eliste and Frederiksson (1998), but this looked only at the impact of economic integration on environmental regulation in the agricultural sector and is, therefore, not directly relevant to the current study.

Bibliography

Agosin, M. and R. Ffrench-Davis (1993) 'Trade Liberalisation in Latin America', *CEPAL Review*, 50.

Antweiler, W., B. Copeland and S. Taylor (1998) *Is Free Trade Good for the Environment?* Vancouver, BC: University of British Columbia, Department of Economics, Discussion Paper No. 98–11.

Barton, J. (1998) 'The North–South Dimension of the Environmental and Cleaner Technology Industries', *CEPAL Review*, 64: 129–50.

Basualdo, E., E. Lifschitz and E. Roca (1988) *Las empresas multinacionales en la ocupación industrial en la Argentina, 1973–1983*. Geneva: International Labour Office.

Birdsall, N. and D. Wheeler (1992) 'Trade Policy and Industrial Pollution in Latin America: Where are the Pollution Havens?', in P. Low (ed.) *International Trade and the Environment*. Washington, DC: World Bank Discussion Papers, 159.

Burki, S. and G. Perry (1997) *The Long March: A Reform Agenda for Latin America and the Caribbean in the Next Decade*. Washington, DC: World Bank.

Cavallo, D. and J. Cottani (1990) 'Argentina', in D. Papageorgiou, M. Michaely and S. Choksi (eds) *Liberalizing Foreign Trade*. Vol. 1. Oxford: Blackwell.

Chudnovsky, D. and A. Lopez (1997) *Las Estrategias de las Empresas Transnacionales en Argentina y Brasil*. Buenos Aires: CENIT, Documentos de Trabajo.

Chudnosvky, D., F. Porta, A. Lopez and M. Chidiak (1996) *Los Límites de la Apertura: Liberalización, reestructuración productiva y medio ambiente*. Buenos Aires: CENIT/Alianza Editorial.

Coes, D. (1991) 'Brazil', in D. Papageorgiou, M. Michaely and S. Choksi (eds) *Liberalizing Foreign Trade*, Vol. 1. Oxford: Blackwell.

Copeland, B. and S. Taylor (1994) 'North–South Trade and the Environment', *Quarterly Journal of Economics*, Vol. CIX, No.3.

Dasgupta, S., H. Hettige and D. Wheeler (1997) *What Improves Environmental Performance? Evidence from Mexican Industry*. Washington, DC: World Bank, Development Research Group Working Papers Series 1877.

Dean, J. (1992) 'Trade and the Environment: A Survey of the Literature', in P. Low (ed.) *International Trade and the Environment*. Washington, DC: World Bank Discussion Papers, 159.

Dijkstra, G. (1997) *Trade Liberalisation and Industrial Development: Theory and Evidence from Latin America*. The Hague: Institute of Social Studies, Working Paper Series No. 255.

Dominguez, L. (1999) 'Comportamiento empresarial hacia el medio ambiente: El caso de la industria manufacturera de la Zona Metropolitana de la Ciudad de México', in A. Mercado (ed.) *Instrumentos Económicos para un Comportamiento Empresarial Favorable al Ambiente en México*. Mexico City: El Colegio de México/Fondo de Cultura Económica.

Edwards, S. (1995) *Crisis and Reform in Latin America: From Despair to Hope*. Oxford: Oxford University Press for the World Bank.

Eliste, P and P. Fredriksson (1998) *Does Open Trade Result in a Race to the Bottom? Cross-Country Evidence*, paper presented to the conference on 'Trade, Global Policy and the Environment'. Washington, DC: World Bank, 21–2 April.

Eskeland, G. and A. Harrison (1997) *Moving to Greener Pastures? Multinationals and the Pollution-haven Hypothesis*. Washington, DC: World Bank, Policy Research Department, Working Paper, 1744.

Ferruntino, M. (1995) *International Trade, Environmental Quality and Public Policy*. Washington, DC: US International Trade Commission, Office of Economics Working Paper.

Fritsch, W. and G. Franco (1991) *Foreign Direct Investment in Brazil: Its Impact on Industrial Restructuring*. Paris: OECD.

Gladwin, T. (1987) 'Environment, Development and Multinational Enterprise', in C. Pearson (ed.) *Multinational Corporations, Environment, and the Third World: Business Matters*. Durham, NC: Duke University Press.

Grossman, G. and A. Krueger (1992) *Environmental Impacts of a North American Free Trade Agreement*. Cambridge, MA. National Bureau of Economic Research, Working Paper No. 3914.

Hogenboom, B. (1998) *Mexico and the NAFTA Environment Debate*. Utrecht: International Books.

IADB (1997) *Economic and Social Progress in Latin America, 1997*. Washington, DC: Inter American Development Bank.

Kosacoff, B. (ed.) (2000) *Corporate Strategies under Structural Adjustment in Argentina*. London: Macmillan.

Lee, H. and D. Roland-Holst (1993) *International Trade and the Transfer of Environmental Costs and Benefits*. Paris: OECD Development Centre, Technical Papers No. 91.

Low, P. (1992) 'Trade Measures and Environmental Quality: the Implications for Mexico's Exports', in P. Low (ed.), *International Trade and the Environment*, Washington, DC: World Bank Discussion Papers, 159.

Mabey, N. and R. McNally (1999) *Foreign Direct Investment and the Environment: From Pollution Havens to Sustainable Development*. London: WWF-UK Report.

Mani, M. and D. Wheeler (1999) 'In Search of Pollution Havens? Dirty Industry in the World Economy, 1960–1995' in P. Fredriksson (ed.) *Trade, Global Policy and The Environment*. Washington, DC: World Bank Discussion Paper No. 402.

Mercado, A. and O. Fernandez (2000) 'Estimación del Volumen e Intensidad de la Contaminación Industrial en México por Entidad Federativa', in S. Puente (ed.) *Medio Ambiente en México*. Mexico, DF: El Colegio de México.

Molina , D. (1993) 'A Comment on Whether Maquiladoras are in Mexico for Low Wages or to Avoid Pollution Abatement Costs', *Journal of Environment and Development*, 2: 1.

Moreno-Brid, J. C. (1999) *Reformas Macroeconómicas y Inversión Manufacturera en México*. Santiago: Economic Commission for Latin America and the Caribbean, Serie Reformas Económicas, 47.

Nordström, H. and S. Vaughan (1999) *Trade and Environment*. Geneva: WTO, Special Studies 4.

OECD (1997) *Economic Globalisation and the Environment*. Paris: OECD.

OECD (1998) *Environmental Performance Reviews: Mexico*. Paris: OECD.

OTA (1992) *Trade and Environment: Conflicts and Opportunities*. Washington, DC: Congress of the United States, Office of Technology Assessment.

Rauscher, M. (1997) *International Trade, Factor Movements, and the Environment*. Oxford: Clarendon Press.

Roberts, T. (1998) *The End of the 'Pollution Haven' as 'Comparative Advantage'? Emerging International Environmental Standards and the Brazilian Chemical Industry*, paper presented at the UMASS–Amherst Conference, 'Space, Place and Nation: Reconstructing Neo-liberalism in the Americas', November.

Rock, D. (1996) 'Pollution Intensity of GDP and Trade Policy: Can the World Bank Be Wrong?', *World Development*, 24(3): 471–9.

Rodrik, D. (1994) 'King Kong meets Godzilla: The World Bank and *The East Asian Miracle*', in A. Fishlow (ed.), *Miracle or Design? Lessons from the East Asian Experience*. Washington, DC: Overseas Development Council.

Ros, J., J. Draisma, N. Lustig and A. Ten Kate (1996) 'Prospects for Growth and the Environment in Mexico in the 1990s', *World Development*, 24(2): 307–24.

Ruijters, Y. (1995) *The Relevance of Environmental Legislation for the Transfer of Environmentally Sound Technology: The Mexican Experience*. Maastricht: UNU/INTECH Discussion Paper #9515.

Runge, C. F., E. Cap, P. Faeth, P. McGinnis, D. Papageorgiou, J. Tobey and R. Housman (1997) *Sustainable Trade Expansion in Latin America and the Caribbean: Analysis and Assessment*. Washington, DC: World Resources Institute.

Schaper, M. (1999) *Impactos ambientales de los cambios en la estructura exportadora en nueve paises de America Latina y el Caribe: 1980–1995*. Santiago de Chile: CEPAL, División de Medio Ambiente y Asentamientos Humanos.

Shaman, D. (1996) *Brazil's Pollution Regulatory Structure and Background*. Washington, DC: World Bank.

Stallings, B. and W. Peres (2000) *Growth Employment and Equity: The Impact of Economic Reforms in Latin America and the Caribbean*. Washington, DC: Brookings Institution.

Ten Kate, A. (1993) *Industrial Development and the Environment in Mexico*. Mexico City: Ministry of Commerce and Industrial Development.

Ten Kate, A. and F. de Mateo Venturini, (1989) 'Apertura comercial y estructura de la protección en México', *Comercio Exterior*, 39: 4.

UNCTAD (1999) 'Trade, Environment and Development, Lessons from Empirical Studies: The Case of Brazil', in *Reconciliation of Environment and Trade Policies: Results of Case Studies*. Geneva: UNCTAD.

UNCTAD (1999) *World Investment Report 1999*. Geneva: United Nations.

UNCTC (1992) *World Investment Report, 1992*. New York: United Nations.

Weeks, J. (1996) 'The NEM and the manufacturing sector in Latin America', in V. Bulmer-Thomas (ed.) *The New Economic Model in Latin America and its Impact on Income Distribution and Poverty*. London: Macmillan.

World Bank (1993) *Malaysia, Managing the Costs of Urban Pollution*. Washington, DC: World Bank.

World Bank (1995) *Argentina Managing Environmental Pollution: Issues and Options*. Washington, DC: World Bank, Environment and Urban Development Division.

World Bank (1996) *Brazil: Managing Environmental Pollution in the State of Rio de Janeiro*. Vol.1. *Policy Report*. Washington, DC: World Bank.

World Bank (1999) *World Development Indicators 1999*. Washington, DC: World Bank.

3

MINING AND THE ENVIRONMENT IN LATIN AMERICA: THE POLLUTION-HAVEN HYPOTHESIS REVISITED

Alyson Warhurst and Nia Hughes-Witcomb

Introduction

The liberalisation of investment regimes over the past decade is accelerating the inflow of foreign direct investment (FDI) into the minerals sector of developing and former centrally planned economies. The FDI currently stands at $300 billion and by far outstrips official flows, which had declined in 1999 to $45 billion.

The environmental effects of FDI in the developing world have been traditionally analysed from the perspective of the pollution-haven hypothesis. This hypothesis suggests that, to the extent that they are spatially footloose in their locational decisions, firms will seek to externalise environmental costs by locating in policy regimes where environmental standards are low and regulations are poorly enforced. This hypothesis, however, tends to downplay the significance of investments in technology and human resources as a result of FDI and assumes that environmental regulations of the host country – or lack thereof – are the most significant determinant of environmental performance.

As its point of departure, this chapter finds that the liberalisation of investment and the privatisation of many state-owned mining companies worldwide is occurring at a time of unprecedented technological and organisational change within the international mining industry. Not only has the decline in ore bodies in the last 10 years increased competitive pressures in the industry, forcing many companies to reassess corporate strategy, but it has also prompted the adoption of more sophisticated technologies and enhanced economies of scale.

Furthermore, the combination of regulatory demands, market opportunities, environmental and social conditions attached to debt, equity investment and political risk insurance are providing incentives for the more dynamic mining firms to invest in the development and/or acquisition of cleaner technologies and management practices. All of these issues lend support to the notion that the effective transfer and assimilation of cleaner technologies and management practices have the potential to improve productivity as well as social and environmental performance at mining operations throughout the developing world.

As such, this chapter will critically analyse the traditional perspective of the pollution-haven hypothesis in the light of these developments, within the context of liberalisation, FDI and enabling environments. Where relevant, it draws on empirical research undertaken within the UK's Economic and Social Research Council's Global Environmental Change Programme.[1]

Mining and investment in Latin America

At the outset, it is important to note that mining is distinct from other industries because ore bodies are neither mobile nor finite. Resource immobility means that companies are not free to choose the most appropriate economic and political climate within which to operate. Mining is a high risk–reward industry, has high sunk costs and requires huge capital expenditure to fund a relatively long development process.

Despite the huge mineralogical potential of Latin American countries, declining mineral prices in the early 1980s and stagnant demand made it increasingly difficult for governments in Latin America, with large, state-owned mining sectors, to finance the management and technical changes necessary to improve competitiveness. Large multinational companies shunned the continent on the basis of the high political and economic uncertainties that dogged the region at this time. This paralysis, coupled with a lack of coherent mining related policies on the part of weak and vulnerable governments heightened business risks to a level that led to the withdrawal of banking interest in most sectors.

Not only did these market conditions preclude reductions in production costs through dynamic development but they also prevented the knock-on benefits of technological changes for lowering externalities.

The decision to undertake fundamental reform often arises as a result of macroeconomic crisis or instability, as the debt crisis of the 1980s aptly illustrates. The concomitant avalanche of new policy prescriptions, in many cases supported by multilateral and bilateral institutions, combined the drivers of stabilisation, liberalisation, open trade and new business environments with falling transport and communication costs and secured inward flows of investment (see Table 3.1). As such, the upward surge in FDI in Latin America has

Table 3.1 FDI flows to selected top 10 developing countries: 1991–7 (billions of US$)

Country	*1991*	*Rank*	*1994*	*Rank*	*1997*[a]	*Rank*
Argentina	2.4	4	3.1	=4	3.8	7
Brazil	1.1	9	3.1	=4	15.8	2
Chile	N/A	N/A	1.8	10	3.5	8
China	4.3	2	33.8	1	37.0	1
Indonesia	1.5	6	2.1	7	5.8	4
Malaysia	4.0	3	4.3	3	4.1	6
Mexico	4.7	1	11.0	2	8.1	3
Peru	N/A	N/A	3.1	=4	N/A	N/A
Venezuela	1.9	5	N/A	N/A	2.9	10
Top 10% share of total FDI[b]	74.2		76.1		72.3	

Source: adapted from Global Development Finance (1998) (World Bank calculations).

Notes

a Preliminary data.

b To all developing countries.

provided a vehicle for economic development and for the transfer of cleaner technologies and production techniques.

It is evident that FDI has increased economic integration into the global market.[2] Cross-border production is taking an increasing share of world output, and the most rapid rise in global production is taking place in developing countries (Lipsey, 1997; Lipsey *et al.*, 1995).

Although developing countries have significantly increased their share of global FDI flows from 20 per cent between 1984 and 1989, to 36 per cent in 1997[3] (*Mining Journal*, 1997b), a significant proportion of these flows has been directed into the mining industry. Total flows of FDI to Latin American countries have been rising due to the continuation of liberalisation, better economic fundamentals and the incidence of large privatisation projects.

Between 1990 and 1995 new mining investment in Latin America totalled $238 billion, while exploration expenditures increased five-fold to $500 million per year, with the devaluation of the Mexican peso in 1994 appearing to have only a short-term effect (Rath, 1995). Capital expenditures on mining alone are estimated to reach $20,000 million for the period 1995–2000 (*Mining Journal*, 1996e).

This trend in exploration expenditure and mining merger activity from the start of the 1990s has provided the impetus for financial markets to meet the demands of the industry. Continued deregulation of capital markets and competitive pressures in the financial sector have prompted innovative and sophisticated mining project finance deals and increased lending levels.

The continued popularity of Latin America as a host for FDI into mining is supported by the fact that the region accounts for 27 per cent of worldwide minerals exploration spending, over twice as much as that flowing to southeast Asia (*Mining Journal*, 1997a). This trend, which has already transcended into higher production levels, should continue as a matter of economics. This is also reflected in the fact that, in 1997, four of the largest five mining project finance deals, in terms of capital cost, were located in Latin America.[4]

The process of reform, both structurally and within the mining sectors of the likes of Chile, Peru and Mexico, has secured impressive inward flows of investment. The extent to which new investments can facilitate the adoption of cleaner technologies, production techniques and improved environmental performance as a result of ongoing liberalisation, restructuring and the availability of finance will be evaluated in the next section, against the traditional perspective of the pollution haven hypothesis.

Pollution-haven hypothesis and mining

The hypothesis

The pollution-haven hypothesis suggests that the liberalisation of investment regimes encourages a spatial displacement of the so-called 'dirty' or pollution-intensive industries away from those countries with stricter environmental regulation and their preferential location in developing countries with minimal environmental regulations or capacity for environmental monitoring and enforcement (see, e.g. Walter, 1982; Leonard, 1987; Richardson and Mutti, 1976).

It also suggests that economic policies which promote foreign investment in the natural resources sector of developing countries will increase rates of natural-resource extraction in the developing world, using conventional pollution-intensive technologies. As such, the theory implies international firms will employ technologies in developing countries that they are no longer permitted to use in the industrialised economies, on environmental grounds, and that may be obsolete with associated energy inefficiencies, etc.

The hypothesis draws its support from two empirical observations. First, that the globalisation of production is associated with increased investment in production in developing countries; and second, that these countries are often characterised by relatively few environmental regulatory provisions and only a nascent capacity for enforcement and monitoring (see, e.g. Gladwin, 1987; Acquah and Boateng, 2000).

This further implies that where foreign investment flows are directed into countries with minimal environmental regulation and a nascent capacity for environmental monitoring and enforcement, then the result will be accelerated environmental degradation (for a rehearsal and critique of this argument, see, e.g. Birdsall and Wheeler, 1992). It then follows that international firms will choose to locate their investment in new capacity in countries that offer the lowest production costs and where there is limited environmental regulation, thereby enabling firms to externalize environmental costs associated with compliance.

Significant high-profile failures, such as the collapse of the tailings dam at the Omai gold operation in Guyana in 1995 and the disputes in the early 1990s over water contamination and forest degradation at BHP's Ok Tedi mine in Papua New Guinea and at Freeport-McMoRan's Grasberg mine in Irian Jaya are cited as evidence that such havens for pollution exist within the international economy (see, e.g. Brown *et al.*, 1993; Bowonder *et al.*, 1985; Bowonder and Linstone, 1987; Shrivastava, 1987; Lagadec, 1987; Pearson, 1987).[5]

However, research into the pollution-haven hypothesis has concentrated almost exclusively on the spatial restructuring of *manufacturing industries* away from industrial economies towards developing countries. This is largely because it is these industries which over the past two decades have been the agents of the 'global shift' in the location of production and have shaped the emergence of a 'new international division of labour' between industrialised and developing countries (see, e.g. Dicken, 1992). It is also because the hypothesis relies on the spatial mobility of capital. It has been implicitly assumed that there is greater mobility of capital in the manufacturing sector than in other sectors, such as primary extraction, because of the constraints imposed by the fixed location of mineral resources and the significance of sunk costs.

What should be taken on board at this point is that, although the location of investments in the minerals sector may be constrained by the location of mineral resources and the richness of the geological deposit, the decision to invest in a specific resource is also influenced by the economic and political conditions relating to investment. Thus, the liberalisation processes abound in developing countries, and the vigour with which the international banking community is responding has opened up unprecedented opportunities for mineral investment in Latin America. As such, these are the drivers that are associated with a switch in the location of new mineral investments away from the industrialised economies, not the weakness of regulatory regimes.

Liberalisation, the pollution-haven hypothesis and environmental performance

There is a range of specific mechanisms through which liberalisation can affect natural resource utilisation and the environmental performance of firms. Although for any given case the specific mix of these mechanisms is largely an empirical question, there is consensus that the impacts of liberalisation on environmental performance can be analysed in one of two ways (Birdsall and Wheeler, 1992). These are the effect of liberalisation on the composition of industry within a specific region – particularly the relative mix of industries of different pollution intensity, known as the 'composition effect' – and the effect of liberalisation on process changes within particular sectors, referred to as the 'process effect'. By analysing the relationships between FDI and regulatory regimes in Latin America and the diffusion and utilisation of better practices, one can contribute to debates on the environmental implications of spatial shifts in the location of minerals investment.

However, much of the evidence generated so far in support of the pollution-haven hypothesis is based on anecdotal evidence in the absence of statistical analysis.[6] As a result, the pollution-haven hypothesis, as conventionally framed, overlooks a number of significant factors by which more open economies, increased FDI and technological collaboration can actually promote improved environmental performance. However, this may be more a result of specific corporate strategy than FDI *per se*.

Technology transfer: the drivers of change

Firms are increasingly being driven to select cleaner processes as part of their package of investments in emerging markets. These drivers include, for example, recently revised and more effective transparent mining and environmental regulations, expectations of increased environmental regulation in the near future and increasing pressures from stakeholders, including financial institutions, to adopt best-practice technologies and techniques (Warhurst, 1996; Warhurst and Bridge, 1997b). A recent trend is the rise in environmental and social conditions attached to the provision of mining project finance and the rigorous environmental and social impact assessments required by multilateral and bilateral financial institutions prior to the allocation of finance. Therefore, cleaner production and better management techniques are implicitly encouraged to meet the requirements of financial institutions (Hughes and Warhurst, 1998).

Indeed, several factors combine to suggest that recent innovations in cleaner process technologies may be diffused as part of a trend towards the globalisation of production in industries such as mining. Research by Romm (1994), Porter and van der Linde (1995), Warhurst (1996) and Warhurst and Bridge (1997a) has demonstrated, for example, that there can be potentially significant competitive benefits for firms that invest in cleaner, less wasteful, and therefore more efficient, production technologies. There can also be significant administrative cost-savings to be gained from standardising corporate environmental practices which reduce the administrative costs of operating in a range of different regulatory environments (Birdsall and Wheeler, 1992).

Liberalisation and regulation

What is strikingly apparent is the way in which the world has become increasingly interdependent economically and financially. In taking up the challenge of competing in demanding and integrated world markets, developing countries have tried to enhance their competitiveness and attractiveness for investment by establishing regulatory frameworks that support market-based policies. As such, the wide range of mineralogical properties now potentially available for investors means that economic liberalisation and associated policies on investment and taxation,[7] in addition to the environment, have a significant role in determining where investment is made.

The liberalisation and stabilisation of economic regimes in Latin America have underpinned the proposition that the growth impact of FDI increases with the openness of the trade regime[8] (Balasubramanyam *et al.*, 1996; Slaughter, 1995). A related study by Wacziarg (1997) also provides evidence to support the dynamic impact that liberalised trade has on GDP growth. He calculated that a one standard deviation increase in an index of trade policy openness is associated with a 0.9 percentage point higher GDP per capita growth. He concluded that the positive impact of openness on growth was due to its effect on investment rates and technology transfer as well as the way in which it induced policy improvements. FDI has also been shown to 'crowd in' more domestic investment, with FDI flows to developing countries being associated with larger increases in overall investment.[9]

Although it is widely accepted that trade liberalisation promotes economic growth, the relationship is hard to prove empirically because of the difficulties in isolating the impact that lower trade barriers have on growth.[10] Economists accept that a country's growth rate is determined by its total factor productivity (TFP). This measures how efficiently a country uses capital and labour. TFP growth can result from domestic and imported innovation. This last is more important in developing countries where there is less human capital. It follows that the speed with which developing countries increase their TFP relative to developed countries will be reflected in their 'openness' (Edwards, 1998).[11]

Therefore, the evidence points to increasing linkages between trade liberalisation, production efficiency and environmental performance.

There is now considerable competition between liberalising countries for foreign mining capital as each country seeks to create a competitive fiscal regime, attractive investment policies and transparent and expeditious permit processing (see Table 3.2).

Table 3.2 Recent policies to promote economic liberalisation in selected developing and transition economies

Country	*Policies of economic liberalisation (South and Central America)*	*Source*
Argentina	Major privatisation programme almost complete, Mining Investments Act, Codigo de Minera, signed in 1993, guaranteeing stability of tax regime for 30 years, capping provincial royalties, and import duty exemptions for mining equipment. Mining Integration Treaty with Chile – June 1996, allowing free flow of mining material and equipment across the border and a five-year plan for the minerals sector from 1995–2000 to promote exploration and development.	MJ 3/5/96 MEG 1993 Hinde 96 MJ 26/4/96 MJ 26/4/96sup MJ 6.3.98
Bolivia	'Capitalizacion' scheme of privatising state industrial assets begun in 1985. Mining and Hydrocarbon Law 1990 improving land access and taxation legislation. More general package of neo-liberal reforms adopted in 1991, including new tax and investment laws allowing 100% foreign ownership and equal treatment of multinationals and domestic mining firms. The revised Mining Code No. 1777 of March 1997 and the Environmental Regulations for Mining Activities of July 1997 provide powerful precedents relating to environmental licensing and regulatory issues	MiAnRe 95 Altamirano 95 MEG 1993 Otto 1995 MJ 6/3/98
Brazil	1996 approval of an updated and streamlined 1968 Mining Code anticipated. Privatisation of state assets initiated 1991, CVRD has opened land holdings to private prospecting in 1995, and privatisation of CVRD in 1997. Tax reduction legislation in 1991 for the repatriation of profits, the ending of discrimination against foreign capital in the mining sector through a constitutional amendment in 1995, an Economic Reform Plan 1994 to tackle inflation and revisions to the Mining Code in 1996.	MJ 24/5/96 MEG 1993 FT 6/6/96 MJ 19/4/96sup Otto 1995 MJ 6/3/98
Colombia	Codigo de Minas 1988	Otto 1995
Cuba	New law allowing foreign investment 1992	Otto 1995
Ecuador	Ministry of Energy and Mines announced planned new mining legislation in 1994. A $24 million World Bank project to improve mining sector, including environmental management	MJ 24/3/95sup
Guyana	Economic recovery programme initiated 1980, private investment in minerals encouraged since 1985, New Mining Act 1989.	MEG 1993 Otto 1995
Mexico	New Mining Investment Law 1989, New Mining Code 1992	Otto 1995

Table 3.2 Continued

Country	*Policies of economic liberalisation (South and Central America)*	*Source*
Nicaragua	Foreign Investment Law 1991, Mining regulatory body established in 1993 to standardise mining activity in the private sector. New Mining Law being drafted (1997) emphasising a reformed role for the state in regulating, promoting and facilitating minerals development	US CIA 1996 Eisen 1998
Peru	Privatisation Decree (Decree 674) 1991, Foreign Investment Statutes (Decrees 662 and 757) in 1991 established guarantees and the legal framework for private investment. Non-discrimination between domestic and foreign investors, provision for tax stabilisation agreements, right to repatriate profits, capital and dividends, guarantees against changes in labour and employment regulations. Supreme Decree in June 1992 revises the General Mining Law to provide a thorough regulatory framework. CENTROMIN included in Private Investment Promotion Process in 1992, but tendering declared void in 1994. Privatisation relaunched 1995 and was finally successful after the government addressed liability issues	Magma 10K Barcellos 1996 MJ 26/1/96 Hughes 1998
Venezuela	Privatisation and neo-liberal reforms, including consolidation of tax code (1994) and new Law of Mining Promotion and Development submitted 1994. A decree passed in April 1996 regarding environmental approval. Two drafts of a Mining Code are presently before Congress. New foreign investment code passed in 1997.	MiAnRe 95 MEG 1993 MJ 6/3/98

Abbreviations: MJ, *Mining Journal*; FT, *Financial Times*.

In Argentina, for example, the Mining Investments Act (Codigo de Minera) was signed in 1993. This covers prospecting, exploration, mineral production and other activities and includes a range of measures which guarantee stability of the municipal, provincial and national tax regime for 30 years, allow 100 per cent income tax deductions for the costs of prospecting and exploration, cap royalties and provide import duty exemptions for mining equipment.[12] In addition to these specific measures aimed at the mining industry, Argentina has passed relatively liberal investment legislation which allows unrestricted transfer of currency overseas and does not preclude the repatriation of capital or the transfer of profits. Subsequently, Argentina has become one of the world's principal targets for exploration expenditure with an anticipated annual total of $135 million by the year 2000 (*Mining*

Journal, 1996c). The story is similar in Bolivia and Chile and, to a lesser extent, in Peru, Colombia and Ecuador.

The race to provide enabling environments has provided competitive pressures that are conducive to the acceleration of investment into more productive technologies and management practices. As such, a growing number of observers conclude that developing countries have benefited from the transfer of cleaner technology and more efficient production techniques[13] (Romm, 1994; Warhurst, 1996; Warhurst and Bridge, 1997a).

Financial drivers

This transformation of investment regimes and of patterns of investment flows is occurring at a time of significant technological change within the mining industry as firms respond to a whole host of pressures. Indeed, the pace of globalisation and deregulation in financial markets coupled with the rise in awareness of sustainability issues are reflected in the way in which mining companies and financial institutions are increasingly addressing social and environmental issues (see the cases of Inti Raymi and Comsur in Bolivia, below).

The incorporation of the goal of 'sustainable development'[14] as a cornerstone of sound business management has been taken up by the financial sector.[15] Financial institutions are increasingly rising to the environmental imperative and attach conditions to project finance and insurance to mitigate the environmental degradation associated with economic development, and therefore protect themselves against future liabilities.

The concept of integrating environmental risks into lending decisions is strategically important in the race to attract new and increasingly environmentally aware clients. This evidence is supported by the fact that over ninety international banks now undertake environmental financial risk assessment of their borrowers and half of these banks incorporate environmental liability into loan terms and monitor environmental risks (Vaughan, 1995). Furthermore, in recognising that environmental risks should be part of the normal checklist of risk assessment and management, President Cardoso passed legislation[16] making it compulsory for Brazilian banks and associated institutions to assess the environmental impacts of any project before committing funds.

Environmental considerations and mining project finance in Bolivia

Inti Raymi S.A. and Comsur S.A. are the largest mining companies currently operating in Bolivia. The IFC (International Finance Corporation, part of the World Bank Group) has played a significant role in providing project finance to both companies since 1989, and other lenders[17] have consistently endorsed IFC requirements with respect to environmental conditions attached to the provision of debt. The credit agreements reached with the IFC, included clauses relating to the borrowers' environmental obligations,[18] are as follows:

> [the borrower should] take all feasible measures to ensure that its business is carried out with due regard to ecological and environmental factors in general and in particular with the World Bank's environmental and safety guidelines. . .

Also,

> [the borrower should] obtain and maintain in force (or where appropriate, promptly renew) all licenses, approvals or consents necessary for the carrying out of the . . . . Project and its business and operations generally; and perform and observe all the conditions and restrictions contained in, or imposed on it by, any such licenses, approvals or consents.[19]

Interestingly, Inti Raymi and Comsur company executives consider that environmental conditions attached to the provision of finance, alongside the environmental policies of their multinational shareholders (RTZ for Comsur and Hemlo Gold and Battle Mountain for Inti Raymi), have been more significant in shaping the environmental behaviour of mining companies than the regulatory framework in place. For example, in 1990, before the Bolivian Environmental Law was passed, RTZ acquired 30 per cent of Comsur's equity. RTZ immediately incorporated much stricter safety and environmental standards into Comsur's operations. In so doing, RTZ brought the project's standards into line with other operations worldwide. Inti Raymi also developed its outstanding Kori-Kollo gold mine in compliance with the World Bank's environmental standards and guidelines and, as noted elsewhere, the mine and processing plant were commissioned while the Bolivian Environmental Law was in the process of being drafted (Loayza and del la Fuente, 1998).

It should also be borne in mind that the involvement of the IFC in many project deals is the *sine qua non* of mine development in many developing countries, and the environmental considerations built into the finance are key drivers of improved performance. Indeed, the role of financial drivers in accelerating the diffusion of cleaner technology and promoting responsible corporate strategy should not be underestimated, especially in developing countries where governments find it difficult to manage the strains of minerals development and, in particular, exchange rate fluctuations. By facilitating FDI flows through the provision of such finance, the international banking community has indirectly promoted the transfer of cleaner technology. It is to this issue that we shall now turn.

Technology transfer and the diffusion of clean technology

Recent research undertaken by the Mining and Energy Research Network (MERN) on the potential of FDI to serve as an effective transfer of technology within the context of liberalisation and at a time of increasing technological change has provided support for the case for the minerals extraction industry.[20]

The diffusion of four clean technologies used in minerals extraction were examined across 25 case studies of technology transfer in Australia, Bolivia, Brazil, Chile, China, Papua New Guinea, Peru, Russia, South Africa and the USA. The principal objective of the project was to examine empirically the conditions under which collaboration between foreign investors and the recipient in the supply of clean technology can enhance environmental management capacity and provide the basis for achieving and sustaining best-practice environmental performance. It assessed the opportunities and constraints for diffusing cleaner production technologies through the mechanisms of FDI.

Some of the key findings of this study are summarised below:

- *Technological collaboration can be an effective means for transferring and developing environmental management capacity in the recipient firm*, but to do so it requires the development in the recipient of mechanisms both to retain capacity and to diffuse it systematically throughout the operation.

> **BOX 1: Morro do Ouro, Brazil**
> Collaboration with a technology supplier to acquire a cleaner process enabled *Morro do Ouro* to meet its principal environmental challenges. At Morro do Ouro in Brazil, the acquisition by Rio Paracatu Mineracao (31 per cent Rio Tinto Brazil, 49 per cent Autram) of the Cyanisorb process for recycling cyanide from the carbon-in-leach (CIL) circuit for the recovery of gold enables the operation to meet and exceed regulatory standards set by the Brazilian environmental agencies.

- *Foreign investment in the mineral sector of developing countries has the potential to serve as an effective vehicle for transferring capacity* as a result of the opportunity it provides for collaboration with a technology supplier to acquire clean technology and the skills and expertise for its operation and improvement. These opportunities, however, need to be purposefully harnessed by supplier and recipient to achieve successful results.

BOX 2: Inti Raymi and Kori Kollo, Bolivia

The acquisition of the INCO SO_2 plant by *Inti Raymi* to destroy residual cyanide in waste at the Kori Kollo gold mining and leaching operation in Saucari province, Bolivia, for example, was regarded as a 'very significant' step towards ensuring best-practice environmental performance. Empresa Minera Inti Raymi, which is 85 per cent owned by US-based Battle Mountain Gold, recovers gold and silver at its 100 per cent owned Kori Kollo deposit located near Oruro, approximately 200 km from La Paz. The site is located in the high desert 4000 m above sea level. In the early 1980s exploration resulted in the discovery of massive oxidized and sulphide gold–silver deposits and in 1982 the Empresa Minera Inti Raymi was founded to exploit these deposits.

Inti Raymi successfully introduced a heap-leaching operation for the exploitation of the oxidised deposit which expanded from 400 to 4000 tonnes per day (tpd) by 1987. In 1993 a 14,500 tpd agitation leaching project was initiated to exploit the sulphide deposit, transforming Kori Kollo into one of the largest gold producers on the continent. *Inti Raymi worked with INCO* to acquire the SO_2/air process for the destruction of residual cyanide from the leach unit. A 20,000 tpd unit was successfully commissioned in 1994. The acquisition of the plant has enabled the company to meet its principal challenge of minimising the potential for groundwater contamination from the tailings disposal facility. The INCO plant was reducing cyanide concentrations in the tailings pond to 15–20 ppm one year before the regulations pursuant to Bolivia's new Environmental Law #1333 were promulgated in December 1995 and two years before the Environmental Regulation of Mining Activities enforced a 50 ppm maximum permissible limitation on weak acid dissociable (WAD) cyanide.

- *Clean technology and best-practice management practices are being adopted as part of foreign investment in the mineral sector of developing and transition economies.* This is particularly the case at greenfield investment sites, but also at those existing sites where there is significant investment in capacity expansion. To the extent that investments in clean technologies are not simple capacity replacements but contribute to expanded mineral throughput and product output, the adoption of clean technologies can be associated with a net increase in local environmental change as a result of expanded mining and processing activity. However, the releases to the environment of potential pollutants can be dramatically reduced as a result of adopting cleaner processes.

BOX 3: Chagres, Chile

Prior to selecting the Outokumpu flash furnace as the technology to assist in capacity expansion in the early 1990s at the *Chagres copper smelter* in Chile, for example, the operating company evaluated potential competitor processes such as Mitsubishi and Isasmelt technology. Operated since 1978 by Compania Minera Disputada de las Condes, a unit of Exxon coal and mineral company, and treating concentrates from the El Salvador and Los Broncos Mines, Chagres approached Outokumpu for a $180 million modernisation and expansion of the smelter from 70,000 to 120,000 tonnes per year (tpy) through installation of an Outokumpu flash furnace, oxygen plant and sulphuric acid plant. The motivation in selecting a new technology was principally economic in that without modernisation of the traditional furnaces and expansion to 120,000 tpy using new technology the company would be unable to compete in the international copper market. Outokumpu was selected as the technology of choice because of its long track record in successfully retrofitting furnaces and reducing costs through reliable technology. The smelter came into operation in January 1995, although a problem with the flash-oven tower resulted in the closing down of the new smelter for a short period. In addition to reducing costs, the smelter improves sulphur recovery and fixation to 95 per cent and allows the plant to meet ambient emission standards while actually increasing copper output. Chagres is the first smelter in Chile to meet international standards for sulphur dioxide.

- *The uptake of clean technology – and the associated opportunities for the improvement of environmental management capacity – is fundamentally dependent on their ability to reduce operating costs.* Uptake typically occurred when environmental advantages could either be translated into cost savings directly – via, for example, reduced energy use or more efficient pollutant capture – or where environmental advantages were incidental to other aspects of the process which rendered it more cost effective. While the successful uptake of clean technology was dependent on the potential cost–benefits, it was also apparent that *clean technology should be considered as a complement to rather than a substitute for regulation.* This accords with other studies (see, e.g. Wheeler and Martin, 1992) which have shown that environmental regulations, by requiring a reduction in environmental releases, can shift the cost curve in favour of clean technologies.

Box 4: El Porco, Bolivia
Similarly, interviews at Comsur's *El Porco* mine in Bolivia demonstrated that the decision of the company to develop an environmental management system (EMS) was fundamental to continue operating at the site. The implementation of an EMS, as a condition of IFC finance, enabled the operation at El Porco to remain in business.

- *Technology suppliers demonstrate enhanced learning in their marketing and collaboration strategies, towards greater emphasis on training and, in particular, emphasis on on-going training.* The objective of these training strategies was to promote skills, understanding and flexibility within the workforce in order to optimise the technical possibilities of the process, particularly during the initial fine-tuning and adaptive phase.
- *The effective use of foreign investment as a tool for transferring capacity to developing countries requires a policy framework to promote innovation and capacity accumulation.* Such a framework is likely to be quite distinct from a more traditional approach of long-term national planning and the provision of protection to key areas of the national economy in order to build up critical technological capacity. There is a need for macroeconomic policy objectives to create incentives for the uptake of cleaner processes and to make sure that cleaner processes are more competitive than existing alternatives. This includes the adoption of clear regulatory standards and timetables recognising the need for technology phase in, as well as policies to encourage innovation and their wider diffusion such as tax breaks for pollution abatement expenditure or research and testing to demonstrate scale-up viability of current bench-level research. In designing policy approaches, however, there is a need to situate technology policies firmly within the context of broader macroeconomic objectives. Policies aimed at innovation, for example, need to recognise other factors within the economy – and other policies – which affect the innovation and diffusion of new technologies. As Correa (1995) and others note, technology policies are generally isolated and poorly articulated with industrial, financial, fiscal, trade and other policies that affect productive activities and competitiveness.
- *With increased foreign investment, the locus of capacity development will change in the developing world from national institutes to private firms, with greater opportunity for implementation.*
- *The development of local capacity can serve as a 'buffer' against competitive or regulatory shocks.* The successful development of indigenous capacity can provide a mineral producer with the abilities to respond adequately to changing market and regulatory conditions. Should prices for a mineral

commodity slide over time, the firm's technical and managerial capacities can be called upon to cut costs, improve productivity and maintain production on site. Similarly, should a new environmental imperative arise in the form, for example, of regulatory requirements to reduce the risk of surface and groundwater contamination, a firm's capacity will enable it to make the modifications to plant and equipment, to management practices and to employee training and development, sufficient to meet and go beyond new regulatory requirements. Faced with changing market and regulatory conditions, capacity enables the producer to buy time in making technological and managerial changes necessary to remain profitable. Similarly, capacity enables a firm to seize new market opportunities rapidly by increasing output or improving environmental performance proactively ahead of regulatory requirements.

These findings do not support the 'pollution haven' hypothesis that suggests firms seek to externalise environmental damage costs by locating in policy regimes where standards are low or poorly enforced. Instead, evidence suggests that productivity and market advantages were obtained by both suppliers and recipients through the efficient diffusion of clean technology. Specifically, the Latin American operations highlighted in the project support the notion that technological collaboration, coupled with FDI mechanisms, in liberalising economies, can improve environmental performance.

Conclusions

The liberalisation of mineral-rich Latin American economies coupled with technological and organisational change have presented the international mining community with unprecedented opportunities. The challenge for Latin American governments is not only to provide enabling environments that encourage investment but also to capture effectively the benefits and manage the environmental performance of mineral resource development.

FDI seems to support the diffusion and/or development of cleaner technology and management practices and these are especially attractive to governments in developing countries, since they hold the promise of reducing environmental damage costs while at the same time maintaining the socio-economic benefits of mining (e.g. resource rents, employment, social infrastructure, skill and technology transfer).

The opportunities for technological and managerial leapfrogging are rapidly expanding as emerging mineral markets design new, transparent mining codes alongside laws that protect the environment; and the conditions attached to the provision of credit, equity investment and risk insurance, promote clean technology acquisition and improved environmental management. The increase in exploration and development programmes in Latin American countries reflects upon the liberalisation efforts and the creation of conducive business environments by their governments.

In this context, greenfield investment and expansions provide opportunities to select state-of-the-art processing technologies from the outset and integrate new production methods with pollution prevention techniques and environmental management systems to achieve both lower cost and environmentally proficient production.

Therefore, the conditions that exist for FDI in Latin America to leverage technological innovation and development are clear. The evidence discounts the traditional perspective of the pollution-haven hypothesis and suggests that the picture is a far more complex one and merits further detailed empirical research. The significance of investments in technology and human resources as a result of FDI should not be underestimated in a rapidly globalising world where companies are being driven to improve their sustainability and responsibilities in their countries of operation.

Notes

1 This paper draws on Professor Warhurst's final report to the ESRC entitled: 'Technology Transfer & the Diffusion of Clean Technology'. A final report to the Global Enviromental Change Programme of the ESRC, 1 November 1995–31 October 1997. See also Warhurst and Bridge (1997).

2 The secular upward trend in FDI flows has not only grown relative to world output, but also relative to international trade.

3 FDI into developing countries totalled $120 billion in 1997 (Global Development Finance, 1998).

4 These were Los Pelambres and El Abra in Chile, SPCC in Peru, and Alumbera in Argentina. The capital cost of these projects totalled US$4,384 million, with a total debt provision of US$2,942 million.

5 It is worth noting that arguments for the existence of pollution havens are not restricted to developing countries, but include allegations of environmental racism and calls for environmental justice in the siting of locally unwanted land uses (LULUs), such as power plants, incinerators or waste disposal sites within the industrialised economies (Bullard, 1993; Gedicks, 1993; Hofrichter and Gibbs, 1993; Pulido, 1993; Pena and Gallegos, 1993).

6 One notable exception is the quantitative analysis undertaken by Eskeland and Harrison (1997). Of the four developing countries tested, they found almost no evidence to support the pollution-haven hypothesis. Their evidence suggests that foreign owned operations in developing countries are less polluting than comparable domestic plants.

7 However, Shah and Slemrod (1991) conclude that special tax advantages and specific fiscal incentives do not significantly influence FDI inflows, yet harmonisation of corporate tax systems to best-practice standards is an important issue.

8 Empirical studies have consistently found a positive and significant relationship between high GDP growth and FDI flows. Indeed, the ratio of FDI to GDP in developing countries has risen from 0.8 per cent in 1990 to 2.0 per cent in 1997 (Global Development Finance, 1998). Nevertheless, causality remains an issue as it may go in both directions. Other studies have also identified the positive relationship between low fiscal and external deficits and high investment flows (Greene and Lillanueva, 1991).
9 Borenzstein *et al.* (1995) found that $1 of FDI led to $0.50 – $1.30 of additional domestic investment.
10 It is important to note that trade can be restricted in many ways, so any measurement of trade liberalisation should not be confined to the impact of tariffs and quotas.
11 Edwards (1998) tested the relationship between trade policy and productivity for 93 countries between 1980 and 1990. In his study he used nine different measures of 'openness' all of which yielded consistent results.
12 The Argentine government also signed a Mining Integration Treaty with Chile in June 1996 to facilitate the free flow of mining materials and equipment in the border area where several major projects and prospects are located (*Mining Journal*, 1996c,d).
13 More specifically, the importance of FDI as a vehicle for the transfer of technology is evidenced in the research of Aitken and Harrison (1994). Basing their research on 4000 industrial facilities in Venezuela, they concluded that foreign equity participation had a positive impact on productivity, with joint ventures consistently outperforming domestic firms.
14 The concept of sustainable development, as defined in the Brundtland Report, was further refined at the UN Conference on Environment and Development in Rio de Janeiro in 1992. The conference put sustainable development firmly on the global agenda and a paradigm shift in the way firms do business. With the adoption of Agenda 21 came the recognition that economic and social development carries with it a responsibility to protect the common global environment, which can be supported by wide-ranging evidence across the stakeholder spectrum. Agenda 21 identifies business and industry as important agents of change, both in the promotion of sustainable development and for economic development. It specifically notes clean technology transfer.
15 The United Nations Environment Programme (UNEP) has been proactive in promoting the environmental imperative. As a result of UNEP bank working-party discussions, a number of leading international banks signed up to the 'Statement by Financial Institutions on the Environment and Sustainable Development' in May 1997. In so doing, signatories officially recognise that '*sustainable development is a corporate commitment and an integral part of good corporate citizenship . . . (and will move) towards the integration of environmental considerations into banking operations and business decisions in a manner which enhances sustainable development*' (UNEP Statement by Banks (1.5), 1992). The role of financial markets in the paradigm shift towards a holistic view of sustainable development is clearly recognised by the World Business Council for Sustainable Development (WBCSD). More to the point, is the understanding that '*unless financial markets can evaluate and reward Eco-Efficiency in business, ... companies will still not move fast enough toward more sustainability.*' (Stigson, 1999).

16 This legislation became commonly referred to as the 'Green Decree'.
17 Other lenders include: DEG: German Development Bank; CAF: Corporación Andina de Fomento – Andean Pact; OPIC: Overseas Private Investment and Citibank.
18 Prior to provision of a loan, the IFC carries out an environmental audit not only at the mine site for which funds are being sought, but also at existing operations. Finally, during the loan period, the borrower authorises the IFC to undertake inspections on the Project's environmental performance.
19 The text belongs to a Credit Agreement between IFC and Comsur.
20 'Technology Transfer and the Diffusion of Clean Technology' The Global Environmental Change Programme of the ESRC.

Bibliography

Acquah, P.C. and A. Boateng (2000) 'Planning for Mine Closure: Case Studies in Ghana' in A. Warhurst and L. Noronha (eds) *Environmental Policy in Mining – Corporate Strategy and Planning for Closure*. Boca Raton, FL: CRC Press.

Agra, G.Y. (1997) *Current Environmental Practice in the Mining Industry*, paper presented at Symposium on Mining and Environment,UST/IDRC Environmental Research Group, Kumasi, Ghana.

Aitken, B. and A. Harrison (1994) 'Do Domestic Firms Benefit from Foreign Direct Investment? Evidence from Panel Data', *World Bank Policy Research Working Paper* 1248.

Altamirano, N. (1996) 'An Environmental Game: The Bolivian Mining Case', School of International Relations and Pacific Studies, University of California, San Diego. (Paper available from the author.)

Andrews, C. (1992) 'Mineral sector technologies: policy implications for developing countries', *Natural Resources Forum*, 16:212–20.

Balasubramanyam, V.N., M. Salisu and D. Sapsford (1996) 'Foreign Direct Investment and Growth in EP and IS Countries', *The Economic Journal*, 106 (January): 92–105.

Barnett, A. (1993) 'The Role of Industrialised Countries in the Transfer of Technology to Improve the Rational Use of Energy in Developing Countries', Science Policy Research Unit, University of Sussex, Brighton.

Birdsall, N. and D. Wheeler (1992) 'Trade Policy and Industrial Pollution in Latin America: Where are the Pollution havens?' in P. Low (ed.) *International Trade and the Environment*. Washington, DC: World Bank Discussion Papers 159:159–68.

Blomstrom, M. and E. Wolff (1989) 'Multinational Corporations and Productivity Convergence in Mexico', Starr Center for Applied Economics, New York University, New York.

Borenzstein, E., J. De Gregorio and Jong-Wha Lee (1995) 'How Does FDI Affect Economic Growth?', Cambridge, MA: National Bureau of Economic Research. NBER Working Paper 5057.

Bouton, L. and M.A. Sumlinski (1996) 'Trends in Private Investment in Developing Countries 1970–1995', Washington, DC: International Finance Corporation, the World Bank Group, IFC Discussion paper, 31.

Bowonder, B. and H. Linstone (1987) 'Notes on the Bhopal Accident: Risk Analysis and Multiple Perspectives', *Technological Forecasting and Social Change*, 32:183–202.

Bowonder, B., J. Kasperson and R. Kasperson (1985) 'Avoiding Future Bhopals', *Environment*, 27:6–13, 31–7.

Brack, D. (1995) 'Balancing Trade and the Environment', *International Affairs*, 71(3):497.

Brown, H. S., P. Derr and O. Renn (eds) (1993) *Corporate Environmentalism in a Global Economy: Societal Values in International Technology Transfer*. Connecticut: Quorum Books.

Bullard, R. (1993) *Confronting Environmental Racism: Voices from the Grassroots*. Boston, MA: South End Press.

Codelco (1998) http://www.codelcochile.com/english/codelco/codelc_f.htm.

Community Aid Abroad (1996) *Annual Review*. Fitzroy, Victoria, Australia: Oxfam.

Correa, C. (1995) 'Innovation and Technology Transfer in Latin America: A Review of Recent Trends and Policies', *International Journal of Technology Management*, 10(7–8):815.

Dahlman, C., B. Ross-Larson and L. Westphal (1985) 'Managing Technological Development: Lessons from the Newly Industrializing Countries', *World Bank Staff Working Papers*, No. 717.

Dechant, K. and B. Altman (1994) 'Environmental Leadership: from Compliance to Competitive Advantage', *The Academy of Management Executives*, 8(3): 7–20.

Dicken, P. (1992) *Global Shift: Industrial Change in a Turbulent World*. London: Harper & Row.

Djankov, S. and B. Hoekman (1997) *Avenues of Technology Transfer: Foreign Linkages and Productivity Change in the Czech Republic*, in Conference on Trade and Technology Diffusion: The Evidence with Implications for Developing Countries, 18–19 April, Fondazione Mattei, Milan, Italy.

E&MJ (1994) 'Mining Opportunities in Latin America', May.

Economist (6.12.97) 'A Survey of Latin America: A Very Big Deal'.

Economist (6.12.97) 'A Survey of Latin America: Back on the Pitch'.

Economist (6.12.97) 'A Survey of Latin America: Buy, Buy, Buy'.

Economist (11.5.98) 'The Perils of Global Capital', 76–8.

Edwards (1998). 'Openness, Productivity and Growth: What do we Really Know?' *Economic Journal*. March.

Eskeland, G.S. and A.E. Harrison, (1997) 'Moving to Greener Pastures? Multinationals and the Pollution Haven Hypothesis', Washington, DC: World Bank, Public Economics Division, Policy Research Department. *Working Paper 1744*.

Financial Times (19.4.95) 'Sustainable Amazon'.

Gedicks, A. (1993) *The New Resource Wars: Native and Environmental Struggles against Multinational Corporations*. Boston, MA: South End Press.

Ghobadian, A., H. Viney, P. James and J. Liu (1995) 'The Influence of Environmental Issues in Strategic Analysis and Choice: A Review of Environmental Strategy among Top UK Corporations', *Management Decision*, 33.

Gladwin, T. (1987) 'Environment, Development and the Multinational Enterprise', in C. Pearson (ed.) *Multinational Corporations, Environment and the Third World*. Durham, NC: Duke University Press.

Global Development Finance (1998) Washington DC: World Bank.

Gooding, K. (1996) 'Globalisation Picks up Pace in the Mining Sector', *Financial Times*, 31 May.

Granstrand, O., E. Bohlin, C. Oskanson and N. Sjoberg (1992) 'External Technology Acquisition in Large Multi-Technology Corporations', *Research and Development Management*, 22(2):111–33.

Greene, J. and D. Lillanueva (1991) 'Private Investment in Developing Countries: An Empirical Analysis', *IMF Staff Papers*, 38.

Hart, S.L. and G. Ahuja (1996) 'Does it Pay to be Green? An Empirical Examination of the Relationship between Emission Reductions and Firm Performance' *Business Strategy and the Environment*, 5 March.

Hofrichter, R. and L. Gibbs. (1993) *Toxic Struggles: the Theory and Practice of Environmental Justice*. Philadelphia: New Society Publishers.

Hughes, N. (1998) 'Economic and Social Policies for Minerals Development in Peru', University of Warwick, UK: *MERN Working Paper* No. 115.

Hughes, N. and A. Warhurst (1998) 'Mining Project Finance', University of Warwick, UK: *MERN Working Paper* No. 151.

Johnson, P.M., A. Beaulieu, V. Lichtinger and M. Johnson (1996) *The Environment and NAFTA: Understanding and Implementing the New Continental Law*. Washington, DC: Island Press.

Johnson, S.D. (1995) 'An Analysis of the Relationship Between Corporate and Economic Performance at the Level of the Firm', *Unpublished Dissertation*. Irvine: University of California.

Kasman, M. (1992) 'Economic and Legal Barriers to the Transfer of Environmentally-Sound Technologies to Developing Countries', *Advanced Technology Assessment System*, 7 (spring):162–70.

Lagadec, P. (1987) 'From Seveso to Mexico to Bhopal: Learning to Cope with Crisis', in P.R. Kleindorfer and H.C. Kunreuther (eds) *Insuring and Managing Hazardous Risks*, Berlin: Springer-Verlag.

Lagos, G. (1997) 'Developing National Mining Policies in Chile: 1974–96', *Resources Policy*, 23 (1/2):51.

Leonard, J. (1987) *Are Environmental Regulations Driving U.S. Industry Overseas?* Washington, DC: Conservation Foundation.

Lipsey, R.E. (1997) 'Globalized Production in World Output', Background paper for *Global Economic Prospects 1997*. Washington, DC: World Bank.

Lipsey, R.E., M. Blomstrom and E. Ramstetter (1995) 'Internationalized Production in World Output', *NBER Working Paper 5385*. Cambridge, MA: National Bureau of Economic Research.

Loayza, F. and J.C. del la Fuente (1998) 'Financial Drivers of Environmental & Social Performance: The Bolivian Case', *MERN Working Paper* No. 160 and part of the final technical report to DfID-sponsored research project ERP206 entitled 'Financial Drivers of Environmental Performance' MERN, Warwick Business School, University of Warwick, UK.

Low, P. (ed.) (1992) 'International Trade and the Environment: An overview', in *International Trade and the Environment*. World Bank Discussion Papers 159: 1–14, 105–20.

Mansfield, E., A. Romeo, M. Schwartz, D. Teece, S. Wagner and P. Brach (1983) 'New Findings in Technology Transfer, Productivity and Economic Policy', *Research Management*, March–April:11–9.

MEG (1993) *Environmental Legislation in the Gold Mining Industry Worldwide*. Halifax, Canada: Metals Economics Group.
Mining Journal (1995) 1 December. CVRD dissected.
Mining Journal (1996a) 15 November, Vol. 327, No. 8404.
Mining Journal (1996b) 6 December, p. 455.
Mining Journal (1996c) 19 April, 'Latin America, Still Emerging'.
Mining Journal (1996d) 26 April, 'Argentina, Exploring the Final Frontier'.
Mining Journal (1996e) 3 May, 'Argentina/Chile Mining Treaty'.
Mining Journal (1997a) 1 August, Finance Supplement. Vol. 329, No. 8440.
Mining Journal (1997b) 12 September, 'America's Still Top'. Vol. 329, No. 8446, p. 1–3.
Mining Journal (1998a) 6 March, 'Investing in Latin America'. Vol. 330, No. 8472.
Mining Journal (1998b) 10 April, Vol. 330. No. 8475, p. 281.
Outokumpu (1998) 'Outokumpu's Statement on the Copper Smelter Project in Northern Chile', Press release from Outokumpu Oyj. 14 January. (Available at www.outokumpu.com/corporat/info/pressrel.nsf/0a94a7c2232e925e422565b4003 1c76f/5625805cd4d3b64225658c004d7f6a?OpenDocument)
Pearson, C. (1987) 'Environmental Standards, Industrial Relocation and Pollution Havens', in C. Pearson (ed.) *Multinational Corporations, Environment and the Third World*. Durham, NC: Duke University Press.
Pena, D. and J. Gallegos (1993) 'Nature and Chicanos in Southern Colorado', in R. Bullard (ed.) *Confronting Environmental Racism: Voices from the Grassroots*. Boston, MA: South End Press.
Porter, M. and C. van der Linde (1995) 'Green and Competitive: Ending the Stalemate', *Harvard Business Review*, Sept–Oct:120–34.
Prast, W. and A. Thomas (1995) 'Mineral investment trends and issues: specific regions', in *Proceedings, International Mining Investment and Regulation Conference*, Centre for Petroleum and Mineral Law and Policy, Dundee.
Pulido, L. (1993) 'Sustainable Development at Ganados del Valle', in R. Bullard (ed.) *Confronting Environmental Racism: Voices from the Grassroots*. Boston, MA: South End Press.
Rath, U. (1995) 'Focus on Latin America', *Mining Journal*, 24 March.
Repetto, R. (1993) *Amicus Journal*, 15(3):34–6.
Richardson, D. and J. Mutti (1976) 'Industrial Displacement through Environmental Controls', in I. Walter (ed.) *Studies in International Environmental Economics*. New York: Wiley, 57–102.
Romm, J. (1994) *Lean and Clean Management: how to boost profits and productivity by reducing pollution*. New York: Kodansha America.
Shah, A. and J. Slemrod (1991) 'Do Taxes Matter for Foreign Direct Investment?', *World Bank Economic Review*, 5(3): 473–91.
Shaw, B. (1991) 'Developing Technological Innovations Within Networks', *Entrepreneurship and Regional Development*, 3(2):111–28.
Shrivastava, P. (1987) *Bhopal: Anatomy of a Crisis*. Cambridge, MA: Ballinger Press.
Shrybman, S. (1990) 'International Trade and the Environment: an Environmental Assessment of the General Agreement on Tariffs and Trade', *The Ecologist*, 20(1):30–4.

Slaughter, M.J. (1995) 'Multinational Corporations, Outsourcing and American Wage Divergence.' *NBER Working Paper 5253.* National Bureau of Economic Research, Cambridge, MA.

Stigson, B. (1999) Speech entitled 'Why Eco-Efficiency? WBCSD Development of the Concept and of Eco-Efficiency Metrics' given at Eco-Efficiency: from Principle to Practice Conference, Sydney, Australia March 15th 1999. Available at www.wbcsd.org/speech/s67.htm accessed September 28th 2000.

Suttill, K. (1991) 'Chile', *E&MJ*, October, 39–43.

Suttill, K. (1994) 'Cuba Turns to the Pragmatists', *E&MJ*, May, 29–39.

Suttill, K. (1996) 'Ecuador', *E&MJ*, April, 31–5.

Vaughan, S. (1995) 'The Greening of Financial Markets', Geneva: UNEP.

Wacziarg, R. (1997) 'Measuring the Dynamic Gains from Trade', background paper for *Global Economic Prospects and Developing Countries.* Washington, DC: International Economics Department, World Bank.

Walter, I. (1982) 'The Relation of International Trade and Environmental Policy', in S.J. Rubin and T.R. Graham (eds) *Environment and Trade.* Totowa, NJ: Allanheld, Osmun, 127–51.

Warhurst, A. (1996) 'The Greening of Minerals Supply – the Cradle', in A. Warhurst and R. Lamming (eds) *The Environment and Purchasing – Problem or Opportunity?* Stamford, Lincolnshire: The Chartered Institute of Purchasing and Supply.

Warhurst, A. and G. Bridge (1996) 'Improving Environmental Performance Through Innovation: Recent Trends in the Mining Industry', *Minerals Engineering* 9(9):907–21.

Warhurst, A. and G. Bridge (1997a) 'Economic Liberalisation, Innovation and Technology Transfer', *Natural Resources Forum*, 21(1):1–12.

Warhurst, A. and G. Bridge (1997b) 'Financing environmental performance strategies', *Mining Finance*, Issue 4:54–56.

Warhurst, A. and R. Isnor (1996) Environmental Issues for Developing Countries Arising from Liberalized Trade in the Mining Industry', *Natural Resources Forum*, 20(1):27–35.

Warhurst, A., P. Mitchell, N. Hughes, K. Franklin, M. Jiwanji, L. Eisen and T. Coleman (1998) 'Social and Economic Policies for Minerals Development in Bolivia, India, Peru, South Africa, Turkey and Zambia', University of Warwick, UK: *MERN Working Paper* No. 115.

Wheeler, D. and P. Martin (1992) 'Prices, Policies and the International Diffusion of Clean Technology: the Case of Wood Pulp Production', in P. Low (ed.) *International Trade and the Environment.* World Bank Discussion Papers 159: 197–224.

Zaelke, D., R. Housman and P. Orbuch (eds) (1993) *Trade and the Environment: Law, Economics and Policy.* Island Press.

4

THE DIFFUSION OF POLLUTION PREVENTION MEASURES IN LDCs: ENVIRONMENTAL MANAGEMENT IN ARGENTINE INDUSTRY

Daniel Chudnovsky, Andrés López and Valeria Freylejer[1]

Introduction

In so far as industrial activities are one of the leading sources of pollution in both developed countries (DCs) and less developed countries (LDCs), it is not surprising that the role played by manufacturing firms in any sustainable development strategy is a critical issue. Whereas in DCs the regulatory framework and the growing concerns of civil society have forced firms to try to reduce the negative impact on the environment of both industrial processes and products, the pressures for incorporating environmental concerns in firms operating in LDCs are not only more recent but also come from different sources.

Although the regulatory framework, its enforcement and the resources allocated to comply with legislation are generally limited in LDCs, the increasing competition in domestic markets due to trade liberalisation has forced firms to reduce costs and to upgrade their technologies to become more competitive. This development has often generated positive environmental effects. At the same time, attempts by firms located in DCs to level the playing field *vis-à-vis* their competitors in LDCs, and the eventual use of environmental measures in OECD markets to constrain access for imports from LDCs, may have induced a greater attention to environmental issues.

According to the traditional viewpoint on the relationship between environmental protection and private costs, environmental regulations can only be met with additional investments and higher operating costs. As a consequence a lower economic growth rate can be expected. Moreover, taking into account the uneven application of environmental regulations in different countries, a trade-off between environmental preservation and private competitiveness – and hence national competitiveness – may often arise.

However, this traditional view has been criticised in different circles. The idea of sustainable development introduced in the 'Brundtland Report' (WCED, 1987) was one of the first studies to put forward the compatibility between environmental protection and economic growth goals. Several authors (Michael Porter among them) have suggested that environmental problems can be substantially reduced, or even avoided, without necessarily incurring higher costs – or at least not unaffordable ones – to the firms. The advantages of pollution prevention (PP) measures against more conventional solutions, e.g. end-of-pipe (EOP) treatments, are often underlined in this alternative viewpoint.

While acknowledging that PP measures are far less diffused than EOP solutions, there is no doubt that it is worth exploring to what extent these approaches are being diffused not only in DCs but also in LDCs. In this connection, the Argentine experience appears as an attractive case to be analysed. Argentina's levels of pollution are more serious than one would expect in a country of middle levels of income per capita and where industrial discharges are a major source of pollution (World Bank, 1995).

Despite this critical environmental background and, in contrast to the significant advances made in other areas of public policy such as privatisation and trade liberalisation, little progress has been made on the environmental front where the regulations are of command and control type and their enforcement is very limited. Furthermore, 'the most critical constraint for improving the management of pollution in Argentina is the absence of clear institutional responsibility for environmental management and the lack of effective enforcement' (World Bank, 1995).

Nevertheless, the environmental situation has slowly and unevenly started to improve due to the pressures of trade liberalisation, the significant inflows of foreign direct investment (FDI), some judicial initiatives and greater environmental concerns among some segments of the population. In some highly polluting sectors, such as pulp and paper, steel, petrochemicals and tanning, we found that firms have started to improve their environmental management (EM) in parallel with their restructuring efforts in a context of greater competition through trade liberalisation, regional integration within Mercosur and growing inflows of FDI (Chudnovsky and Chidiak, 1996; Chudnovsky *et al.*, 1996). Similar improvements in EM have been reported in other studies made in the pharmaceutical and food-processing industries, though the motivations differ in each sector (FIEL, 1996).

Against this background, the main objective of this chapter is to gain a deeper knowledge of Argentine industrial firms' EM, especially of SMEs of which almost nothing is known, and to study the progress made in the adoption of PP measures by addressing the following issues:

1 The obstacles faced and the motivating factors for adopting the PP approaches and the type of measures most frequently adopted by local firms;
2 The sources of the technologies adopted for implementing PP measures and the role played by endogenous technological capabilities within the adopting firm;
3 The economic consequences of the adoption of PP projects in the Argentine case.

Our expectation was that a more active EM and a wider adoption of PP measures would have taken place in larger, export oriented, foreign owned firms and/or in more technologically innovative firms. Furthermore, enterprises operating in high-pollution potential sectors would generally have had a similar behaviour.

Given the absence of information, detailed questionnaires were prepared and two surveys were carried out between the end of 1996 and mid-1997.[2] Their findings are the main source of information for shedding light on these issues. Whenever possible, the findings are compared with previous information on the Argentine industry and with similar studies carried out in other Latin American countries.

In the next section the potentials of the PP approach are summarised, and in the following section the opportunities and constraints in adopting PP measures, especially in LDCs, are analysed. The empirical evidence about Argentine firms' EM, focused on the diffusion of PP measures, is then examined. The concluding remarks, further research suggestions and policy implications are dealt with in the final section.

Potential of a PP approach[3]

The traditional viewpoint on the relationship between environmental protection and private costs is based on the static way of thinking, in which technology and customer needs are all considered as given, information is perfect and profitable opportunities for innovation have already been discovered. Instead, the actual process of dynamic competition is characterised by changing technological opportunities coupled with highly incomplete information and organisational inertia. In this context, properly designed environmental standards can trigger innovation that may partially or fully offset the costs of complying with them. Such 'innovation offsets' – broadly divided into product and process offsets,[4] as Porter and van der Linde (1995a) call them, would not only reduce pollution, but improve the productivity with which resources are used.

> Ultimately companies and regulators must learn to frame environmental improvement in terms of resource productivity. At the level

> of resource productivity environmental improvement and competitiveness come together. In this connection, firms can benefit from properly designed environmental regulations. By stimulating innovation, strict environmental regulations would enhance competitiveness and reduce pollution levels leading to a 'win–win situation'.
> (Porter and van der Linde, 1995a,b)

Although this ultra-optimism is not entirely shared among all the proponents of this viewpoint, the advantages of PP measures against more conventional solutions – EOP[5,6] – are unanimously underlined. The crucial idea is to shift from a 'corrective' approach to a 'preventive' one in EM. The development of an innovatory capability to find preventive solutions for pollution problems in the productive sector becomes a key element in making this fundamental change possible.

Compared to the conventional treatment alone, PP and recycling investments are often more cost effective. PP may produce significant environmental benefits as well, including reduced cross-media transfers and reduced environmental impacts from avoided energy and materials usage. However, while increased reliance on PP and recycling offers a means to reduce the conflict between environmental protection and industrial competitiveness, it does not eliminate it. While many PP and recycling options yield net positive rates of return equalling non-environmental investments, many others do not and often cost money. However, in most cases the expense is lower than alternative EOP approaches (OTA, 1994).

PP measures normally include: good housekeeping, maintenance and operating practices; product reformulation and raw material substitution; relatively simple process modifications employing currently available technologies; and more fundamental process modifications, mainly requiring technological innovation and external recycling (OTA, 1994).

PP actions should be distinguished according to their level of complexity. There are some 'simpler' ones – with small investment requirements, low technological complexity and short implementation periods, such as water, energy and input savings. At the other end, there are more 'complex' measures, generally involving greater investments, longer lead times and higher technological complexity and uncertainty. The development of new cleaner technologies is a good example of more 'complex' actions.

At the same time, there are many similarities between PP and total quality management (TQM), a key instrument for competing in open economies.

> In both firms examine their production process in great detail and focus on continually improving the process to improve quality and productivity and reduce scrap and pollution. Both practices incorporate new cost accounting and measurement to assign all costs to particular products or production processes. Benchmarking progress

> is encouraged in both. In TQM, firms strive for zero defects, while in the best pollution prevention efforts, firms strive for zero discharges. . . Both practices aim to involve all parts of a company. . . Similarly both stress the importance of workforce involvement and the key role of shop-floor workers in improving quality and preventing pollution.
>
> (OTA, 1994).

PP in practice

The PP approach has been encouraged in several DCs. As a consequence, manufacturing firms in these countries have been adopting it, motivated by the idea of finding cheaper solutions to face their EM. However, these measures (especially the more 'complex' ones) have not yet been widely diffused.

According to Hanrahan (1995), exogenous and endogenous factors to the firms account for such a situation. An appropriate regulatory framework appears as a prerequisite for a wider diffusion of PP programmes. The internal dynamics, the managerial, productive and innovative capabilities and the sector of operation are some of the factors accounting for the adoption of PP approaches within firms. Moreover, lack of information on specific pollution levels in each firm may mean that PP opportunities are sometimes foregone. In addition, the lack of appropriate information on the availability and costs of PP options constrain the diffusion of PP measures, especially among SMEs (OTA, 1994). Supply restraints, such as the non-existence of the necessary technologies or the presence of an environmental lock-in situation where the equipment producers are 'biased' to EOP solutions, could also explain the limited diffusion of the PP approach.

In view of the current state of the environmental protection debate, the obstacles to a wider diffusion of PP methods should necessarily be faced not only in DCs but also in LDCs. In the case of LDCs, the impact of environmental regulations on economic growth and competitiveness is increasingly becoming a central concern. In particular, the implementation of voluntary measures – ecological labels or ISO 14000 certifications – and the eventual restrictions because of 'ecological dumping' – which represent real or potential threats to the competitiveness of exports from LDCs to DCs – are being gradually considered.

Although in LDCs environmental regulations are seldom enforced and firms do not often have an endogenous capability to absorb and adapt modern technologies, it is possible that this situation has begun to change. The opening up of the economies, the resumption of economic growth, pressures from clients in DCs, the increasing attention that TNCs are giving to environmental issues – both in the parent companies and affiliates – and growing preoccupation with the local environment are the new incentives which should stimulate the adoption of new practices and/or technologies that reduce both private costs and pollution levels.

However, not all firms react in the same manner and with similar technological responses to the new stimuli. The different possible paths open to the firm depend on the nature of its accumulated competencies and learning skills; in other words, its evolutionary direction is predetermined by the nature of its specific assets (it is path dependent). Firms which have the resources and capacity to innovate and to harness technological and organisational change should be in the best position to face the challenge. The possibility of developing innovatory and technological capabilities would depend on the size and nature of the firm, its sector of operation and its accumulated intangible assets.

In fact, since many firms, especially in LDCs, have limited technological and organisational capabilities, once the environmental dimension appears as a new variable to them, they tend to choose EOP solutions. These generally do not lead to firms changing their production methods, global organisation or organisational and management structures. In contrast, more 'complex' PP measures are harder to implement; they demand planning, design, production and marketing activities' redefinition, as well as global management reorganisation, in order to include environmental concerns in each one of these stages. Hence, in view of organisational inertia, uncertainty about innovations and limited learning skills, it may be expected that the adoption of PP measures, excepting 'simpler' ones, would be low and gradual.

The received literature on technological innovation has shown that firms in LDCs have seldom developed major innovation capabilities. Nevertheless, despite this deficiency, some production and investment capabilities have been created during the long process of import substitution industrialisation, especially in continuous process branches like steel, oil refining, petrochemicals, pulp and paper and food processing, which are some of the most polluting branches in the manufacturing sector. With such capabilities firms are normally able to carry out process and product engineering activities like equipment stretching, process adaptation, reduced downtime, energy and raw materials savings and product quality improvement. However, firms generally rely on different foreign sources for new technologies and sophisticated capital goods.

Regarding technologies owned by TNCs, these firms would probably influence the environmental performance of their affiliates' suppliers, competitors and/or customers, both by 'demonstration effect' and by the introduction of their own environmental standards. Furthermore, local staff training on pollution control technologies, waste minimisation and dangerous waste handling may also be provided by the parent companies (O'Connor and Turnham, 1992; UNCTAD, 1993).

The spillovers from TNC operations would also depend on the accumulated productive and technological capabilities of domestic firms, on the available infrastructure and amount of qualified human resources and on the receiving country's public policies. Furthermore, as technical knowledge is partly tacit and localized, to master the imported technologies firms in LDCs have to develop an endogenous capacity to absorb, adapt and modify new technologies (Dahlman *et al.*, 1987; Katz, 1990; Lall, 1992). For these purposes, firms have to employ skilled technical staff and train their manpower in a systematic way, hence contributing to human resource development in the country.

It is likely in this context that firms with such technological capabilities would be in a better position to absorb and adapt PP and waste minimisation technologies that are available at the international level, through different channels. Furthermore, since pollution problems are very location specific, an endogenous technological capacity to find *ad hoc* solutions to these problems should normally be required.

Diagnosis of the EM situation in Argentine firms: an heterogeneous picture

Diffusion of PP measures within industrial firms' EM

Before focusing our attention on the surveys' main findings, it is important to make a clarification. Due to the lack of public information on the amount and type of emissions, at sectoral or type of firm levels, it is not possible to examine the environmental performance of the firms as such. Therefore, attention is paid to the firms' EM. Though it is not always true that a better EM represents a superior environmental performance,[7] in general one can assume that firms with a more active EM have the conditions to control their pollutant emissions in a more efficient way than those with poor EM. Certainly, a firm with an active EM would probably achieve a better environmental performance per product unit than their competitors in the same sector with a weaker EM. At the same time, after adopting such an EM approach the environmental performance of the firm should improve.

In previous studies made by CENIT we found that firms have started to improve their EM in parallel with their restructuring efforts. A better environmental performance has often been achieved as a by-product of the efforts made to reduce costs and increase production efficiency to face the growing competition in domestic and export markets.

While EOP treatment has been unevenly incorporated in many facilities, there is limited evidence of PP and waste minimisation activities. In this connection, several firms have undertaken process optimisation and waste re-use activities as part of their efforts to reduce costs and save energy (whose prices are now at international levels) and only a few firms have adopted prevention technologies, especially in the pulp and steel sectors (Chudnovsky and Chidiak, 1996; Chudnovsky *et al.* 1996).

These findings are strengthened by the results of the studies carried out two years later. In fact, some progress is visible in local firms' EM, especially among larger ones. More than 90 per cent of surveyed large firms have an environmental department. In addition, they have defined their own environmental policies and have also established targets for their environmental performance. Moreover, all these firms have staff exclusively assigned to EM although not often full-time personnel. These enterprises are aware of local and international environmental regulations and, in several cases, they are members of business associations created for the purpose of diffusing new EM standards. Although their EM is behind the international best practices, most of them have adopted primary and secondary treatment facilities, or similar EOP facilities.

In addition, regarding environmental investments, the available evidence reveals a growing trend between 1993 and 1997. Moreover, the average share of environmental investments in total investments has been increasing since 1993 (Table 4.1).[8]

In contrast, surveyed SMEs have shown strong deficiencies in their EM. Correspondingly, it is very difficult to expect that the SMEs would have an environmental unit and/or assign specific personnel for dealing with environmental issues. Meanwhile, only 20 per cent of the surveyed SMEs have EOP facilities. Likewise, most of the SMEs (60 per cent) do not have environmental performance targets and, in several cases, know neither the current national nor the provincial environmental regulations.

Other Latin American studies confirm the perception of a weak EM among SMEs. Some examples are mentioned in a study carried out in the Metropolitan Zone of Mexico City[9] (Domínguez-Villalobos, 1999) and in other studies included in Brugger *et al.* (1996).[10] Coming back to large Argentine firms, the available evidence indicates that 92 per cent of surveyed firms carried out environmental training of their staff (even though the time devoted to environmental training activities and the number of persons

Table 4.1 Large firms' environmental investments: 1993–7 (thousands of US$ and percentages)

	1993	*1994*	*1995*	*1996*	*1997*[b]
Environmental investments (on average)[a]	1877	2163	2209	1504	3054
Environmental investment in total investments (on average)[a]	10.0	11.9	8.3	18.7	19.0

Notes
a Total number of firms providing information on each item is not always the same.
b Estimated.

involved are not very significant yet). Moreover, it is worth highlighting that almost 13 per cent of the surveyed firms have achieved one environmental certification (ISO 14000), whereas another 40 per cent have at least one in progress.

The implementation of an environmental accounting system is another development taking place at several firms (almost 40 per cent of surveyed large firms). Besides the logical interest in learning the EM costs, such efforts may facilitate a cost–benefit analysis of the different EM options and would probably encourage the adoption of PP measures. In fact, PP practices have increasingly been adopted by surveyed large firms as part of their EM. As may be expected, adopted measures are those which are 'simpler' (energy, water and input savings,[11] followed by good housekeeping, maintenance and operating practices and by staff training). More 'complex' measures (such as process modifications, cleaner new technology adoptions, raw material substitution and product reformulation) have been less important.

The importance given to PP measures in the EM of the large firms surveyed is also illustrated by the information available regarding the share of PP investments in their overall environmental investments – they accounted for 50/60 per cent in the period 1994–7.

These findings are in line with those described in Chudnovsky *et al.* (1996). In those studies it was clear that environmental practices leading to a positive economic return were adopted first. Hence, measures demanding low-cost investments and leading to recovery of raw materials and/or by-products had been the most diffused. The limited investment efforts in new facilities or production lines embodying cleaner technologies, except in connection with the adoption of international standards, were also found in our previous study.

A similar bias in favour of cost-effective PP measures have also been revealed in other Latin American countries. In Mexico, several enterprises completed programmes to increase the efficiency of water, power and fuel use. A trend towards the adoption of recycling methods was also visible (Domínguez-Villalobos, 1999). In contrast, in Chile, clean technology facilities were more diffused than in the rest of the region, e.g. in the pulp and paper sector, linked to the recent significant industrial expansion of this branch (Scholz *et al.*, 1994). Meanwhile, the adoption of PP measures among SMEs has a similar pattern, even though there is much less diffusion than in large firms.

An interesting point to highlight among large firms is the difference in EM (and particularly in the adoption of PP measures),[12] according to their market orientation. As shown in Table 4.2, EM is stronger in export-oriented firms than in those geared to the domestic market. Besides, quality management standards and, more recently, EM ones (such as ISO 9000 and ISO 14000) are important factors among export-oriented firms. Most export firms

Table 4.2 Large firms' environmental management according to market destination (percentages)

	Weak EM	*Medium EM*	*Active EM*	*Total*
Export oriented (17 firms)	12	35	53	100
Internal-market oriented (15 firms)	27	53	20	100

surveyed have already obtained an ISO 9000 certification while only 25 per cent of the firms selling to the domestic market have the same certification. Regarding ISO 14000 certifications this tendency is also visible, even though it is still incipient.

Regarding the diffusion of PP measures, a weak PP management has more often been found in firms mostly selling to the domestic market than in export-oriented ones (Table 4.3).

As assumed, there is also a difference in EM between foreign and domestic enterprises. Broadly speaking, EM is more advanced in foreign firms than in domestic ones. Nevertheless, there is also a greater percentage of foreign firms with a weak EM (Table 4.4). In the specific case of TNCs with a weaker EM it is not likely that they are behaving as 'environmental refugees'; in fact, they are firms that have recently been purchased by foreign investors and where environmental practices have been inherited from preceding local owners. Meanwhile, most foreign enterprises with an active EM apply the global policies defined by their headquarters, even though, in some cases, the subsidiaries keep some autonomy to react to specific local circumstances.

Furthermore, on average, PP measures have been adopted more by TNCs subsidiaries than by domestic firms (Table 4.5).

Finally, an annoying feature is that those firms operating in high-pollution potential sectors not only display on average a weaker EM than the rest of the surveyed firms, they have also made little progress in the adoption of PP measures (Tables 4.6 and 4.7). Besides the threat of serious environmental problems, this finding suggests that possible solutions to these problems may be beyond local firms' present capabilities and resources.

Table 4.3 Large firms' adoption of PP practices according to market destination (percentages)

	Weak PP management	*Medium PP management*	*Active PP management*	*Total*
Export oriented (17 firms)	18	47	35	100
Internal-market oriented (15 firms)	40	13	47	100

Table 4.4 Large firms' environmental management according to origin of the firm (percentages)

	Weak EM	*Medium EM*	*Active EM*	*Total*
Domestic (15 firms)	14	57	29	100
Foreign (17 firms)	18	35	47	100

Table 4.5 Large firms' adoption of PP practices according to origin of the firm (percentages)

	Weak PP management	*Medium PP management*	*Active PP management*	*Total*
Domestic (15 firms)	29	50	21	100
Foreign (17 firms)	23.5	17.5	59	100

Table 4.6 Large firms' environmental management according to firms' pollution potential (percentages)

Sectors according to pollution potential	*Weak EM*	*Medium EM*	*Active EM*	*Total*
Low (12 firms)	8	50	42	100
Medium (4 firms)	0	25	75	100
High (16 firms)	31	44	25	100

Table 4.7 Large firms' adoption of PP practices according to firms' pollution potential (percentages)

Sectors according to pollution potential	*Weak PP management*	*Medium PP management*	*Active PP management*	*Total*
Low (12 firms)	17	25	58	100
Medium (4 firms)	0	50	50	100
High (16 firms)	44	31	25	100

Obstacles and motivating factors for the adoption of the PP approach

Regarding the obstacles for the adoption of PP measures, the access to cleaner technologies is the main one faced by the firms, both foreign and domestic. Although this barrier is more frequent among domestic enterprises, it is also significant among TNC subsidiaries. Furthermore, the lack of monetary and/or human resources also constrains the adoption of PP measures, especially among domestic firms.

The access to cleaner technologies is also the principal restraint preventing adoption of PP measures among SMEs (where PP measures are, as mentioned, far less diffused). In addition, the lack of information is an obstacle for some of the surveyed SMEs. Another crucial finding is that more than 25 per cent of SMEs have not been able to point out which are the difficulties which prevent them improving their EM.

The reduction of EM costs is the main motivating factor in the adoption of PP measures among large firms, followed by the willingness to improve the firm's 'environmental image'. Operating cost reduction actions (leading to PP as a by-product) is the third factor encouraging the diffusion of PP measures. National regulations appear in fourth place (since they are of command and control type, it is difficult to expect that they may have an important influence). Preparatory work to obtain environmental certification also appears as a motivating factor. In contrast, domestic and external market demands, financial institutions' requirements and the emulation of local competitors' actions do not appear as important factors.

However, all the factors mentioned behave in a different way according to the type of measures and the kind of firms surveyed. For example, the enhancement of the 'environmental image' is the main stimulus for the adoption of product reformulation, customer/supplier co-operation and staff-training activities. Likewise, water, energy and input savings, raw material substitution or external recycling are mainly fostered by the need to reduce EM costs and/or they come forth as by-products of actions aimed at reducing the operating costs of the firm.

Another finding is that EM cost reduction is the main stimulus to adopting PP measures among export-oriented firms, followed by the enhancement of the firms' 'environmental image', national environmental regulations and preparation to obtain environmental certifications, which reflect the need to fulfil external requirements. However, external market demands have not played an important role in the adoption of PP measures among these firms.

Finally, the enhancement of firms' 'environmental image' is the main motivating factor among foreign enterprises, followed by the EM cost reductions. Also, PP practices arise in several cases as a by-product of actions aimed at reducing the operating costs of foreign firms. Otherwise, local environmental regulations do not appear as a very significant factor for the adoption of PP measures, reflecting the fact that several subsidiaries may be implementing their own standards, perhaps higher than those in force in Argentina.

Innovation, technology sources and preventive management

An important finding is that in-house activities are the main source for the technologies required to adopt PP measures, for those measures which are easy to implement and/or where the problems are firm specific. Not

Table 4.8 Large firms' environmental management according to firms' innovative capabilities (percentages)

	Weak EM	*Medium EM*	*Active EM*	*Total*
Low (10 firms)	30	60	10	100
Medium (10 firms)	20	30	50	100
High (12 firms)	8	42	50	100

surprisingly, other sources are as important as in-house activities when product reformulation or the adoption of a cleaner technology is required. Presumably, the headquarters appear as a substantial source of technology for foreign enterprises. Besides that key source, TNCs affiliates rely for their technological inputs also on specialised local enterprises (instead of foreign ones) and to some extent on local universities and/or research institutes.

In contrast, besides the role of in-house activities as the main source of technology, domestic firms also rely on specialised foreign firms[13] (instead of domestic ones) and they have few links with local universities and/or research institutes. Furthermore, a positive relationship between innovatory capabilities/quality management and EM and adoption of PP measures has been found (Tables 4.8 and 4.9). These findings are in line with similar ones mentioned in other Latin American studies.[14] This is a clear reflection of the importance of endogenous technological capabilities in the development of a proper EM. Moreover, the capabilities required for good quality management are related to those associated with good EM. As mentioned above, there is a convergence between TQM and PP measures.

Likewise, firms operating in medium/high and high technological content branches[15] display a better EM and have made more progress in the adoption of PP practices (Tables 4.10 and 4.11). At an international level these branches are the most dynamic and innovative ones. Hence, they have more possibilities of developing innovative environmental solutions from which local firms may also benefit.

Table 4.9 Large firms' PP management according to firms' innovative capabilities (percentages)

	Weak PP management	*Medium PP management*	*Active PP management*	*Total*
Low (10 firms)	50	30	20	100
Medium (10 firms)	20	30	50	100
High (12 firms)	17	33	50	100

Table 4.10 Large firms' environmental management according to sectoral technological contents (percentages)

Sectors according to technological content	*Weak EM*	*Medium EM*	*Active EM*	*Total*
Low (11 firms)	27	62	11	100
Medium–low (7 firms)	28.5	28.5	43	100
Medium–high (7 firms)	0	14	86	100
High (3 firms)	0	67	33	100

Table 4.11 Large firms' PP management according to sectoral technological contents (percentages)

Sectors according to technological content	*Weak PP management*	*Medium PP management*	*Active PP management*	*Total*
Low (11 firms)	27	55	18	100
Medium–low (7 firms)	43	43	14	100
Medium–high (7 firms)	0	0	100	100
High (3 firms)	33	0	67	100

Economic advantages of the adoption of PP measures

An interesting finding is that the adoption of PP measures, that in general do not eliminate the need for keeping EOP facilities, has meant economic advantages with respect to more traditional control methods. In fact, 70 important PP projects carried out between 1992 and 1997 have been identified in large surveyed firms. Only in 17 per cent of the projects have the expenditures for adopting PP measures involved a non-recovered cost. The expenditures made have been at least partially recuperated in the remaining projects. In more than 20 per cent of the projects additional monetary benefits have been obtained.

Firms with a better EM, more advanced adoption of PP practices and higher innovatory and quality capabilities have developed most PP projects and especially the most profitable ones (Tables 4.12–4.14). Note that none of the PP projects with totally recuperated costs or with net benefits have been developed by firms with a weak EM. Meanwhile, most of these types of projects have been made by enterprises with medium/high innovatory and quality capabilities and making more progress regarding PP methods (Tables 4.12–4.14).

Another finding is the weak relationship between the origin of the firm and the project results. Since foreign firms have advantages in obtaining new technologies and innovations concerning PP measures, they would be in a better position to develop projects with greater economic and environmental results. However, it does not seem to be the case in our sample (Table 4.15).

Table 4.12 Large firms' 'preventive projects' results according to firms' environmental management

Firms' EM	*Expenditures for adopting PP measures*								
	NRC[a]		*PR*[b]		*CTR*[c]		*Total of projects*		*Number of projects per firm (on average)*
	Number	*%*	*Number*	*%*	*Number*	*%*	*Number*	*%*	
Weak	4	33	5	18	0	0	9	13	1.5
Medium	6	50	12	43	12	40	30	43	2.1
Active	2	17	11	39	18	60	31	44	2.6
Total	12	100	28	100	30	100	70	100	2.2

Notes
a No recuperated cost.
b Partially recuperated cost.
c Totally recuperated cost; in some cases jointly with additional benefits with a similar or lower rate of return than other non-environmental investments.

Table 4.13 Large firms' 'preventive projects' results according to firms' PP management

Firms' PP management	*Expenditures for adopting PP measures*								
	NRC		*CPR*		*CTR*		*Total of projects*		*Number per firm*
	Number	*%*	*Number*	*%*	*Number*	*%*	*Number*	*%*	
Weak	4	33.3	8	28.5	5	17	17	24	1.9
Medium	4	33.3	8	28.5	11	36	23	33	2.3
Active	4	33.3	12	43	14	47	30	43	2.3
Total	12	100	28	100	30	100	70	100	2.2

Table 4.14 Large firms' 'preventive projects' results according to firms' innovative capabilities

	Expenditures for adopting PP measures								
	NRC		*CPR*		*CTR*		*Total of projects*		*Number per firm*
	Number	*%*	*Number*	*%*	*Number*	*%*	*Number*	*%*	
Low	3	25	11	39	2	6.6	16	23	1.6
Medium	4	33	6	22	14	46.6	24	34	2.4
High	5	42	11	39	14	46.6	30	43	2.4
Total	12	100	28	100	30	100	70	100	2.2

Table 4.15 Large firms' 'preventive projects' results according to origin of the firm

Firms	*Expenditures for adopting PP measures*							
	NRC		*CPR*		*CTR*		*Total of projects*	
	Number	*%*	*Number*	*%*	*Number*	*%*	*Number*	*%*
Domestic	5	42	12	43	15	50	32	46
Foreign	7	58	16	57	15	50	38	54
Total of projects	12	100	28	100	30	100	70	100

Meanwhile, for the existing EOP treatments the adoption of PP measures in each firm has an uneven impact. For more than 57 per cent of the firms, such adoption has allowed them to reduce the operating costs of existing EOP facilities. Furthermore, the investment needs in the expansion and/or incorporation of EOP treatments have decreased in several cases (50 per cent). Finally, a small number of firms (35 per cent) have modified their productive equipment with the aim of eliminating the use of EOP facilities.

Concluding remarks and policy implications

In a context of incipient improvement in the EM within the Argentine manufacturing sector, the progress made is, as expected, different according to the size, origin of capital and market orientation of each enterprise. Argentine firms' EM evolution in the 1990s has been affected by factors such as local environmental regulations and pressures, external market demands, TNC strategies and changes in competitive conditions in the local market. Since each firm has its own inherent characteristics and also its own view of the relationship between EM and competitiveness, the scope of EM is very uneven; furthermore, the factors mentioned do not lead to the same effects in each surveyed firm.

As expected, PP approaches are more diffused among large firms, particularly those with higher export orientation and/or which are foreign controlled. In contrast, the lack of knowledge on overall environmental matters is widespread among SMEs. The PP measures adopted are those demanding small investments and/or costs and requiring relatively 'simple' technological actions.

Another important finding is the positive relationship between innovatory/quality capabilities and EM and adoption of PP practices. Moreover, in-house activities have been of great significance in the adoption of PP measures. Nevertheless, since only 'simple' PP practices have been adopted, the positive connection between innovatory capability and EM – particularly the adoption of PP measures – seems to take place within narrow bounds. In-house activities have been the main source of technology for the adoption of

PP measures, even among TNCs subsidiaries, which anyway depend on technology flows from their headquarters. This finding is related to the fact that the more diffused measures generally are firm specific. Hence, internal staff are able to develop appropriate solutions linked to their own experience and learning (i.e., it is a tacit and specific knowledge). However, endogenous efforts are also important, though with less intensity when more technologically 'complex' measures such as the adoption of cleaner technologies and/or product reformulation are undertaken.

In fact, the access to more 'complex' technologies appears as the main obstacle to making more progress in the adoption of PP practices, both in SMEs and in large firms. Further research is needed to interpret this finding. In large firms it may reflect that insufficient resources have so far been allocated to the adoption or development of environmental technologies. It may also reflect the poor investment record in new facilities and machinery embodying such technologies. It is also possible that cleaner technologies are not yet available even in DCs.

Economic reasons – EM cost reduction – seem to be the leading motivating factors for adopting PP practices. A surprising finding is the great importance given to enhancing 'environmental image' as a stimulus to adopting PP measures, particularly among TNC subsidiaries. Furthermore, it has been common that PP measures appear as by-products of actions aimed at reducing the operating costs of the firm. National regulations have not played a significant role (though domestic firms consider them to be more important than do foreign ones).

In contrast, external market requirements have not been so relevant for the adoption of PP solutions, even for export-oriented firms. Although this type of firm has made more progress in the adoption of PP practices, it seems that it is motivated more by the need to reduce cost to compete in international markets rather than by specific requirements of such markets. Nevertheless, this finding should not be taken as conclusive, especially considering that according to other studies carried out by CENIT the pressures from customers in export markets have had an important role both in the adoption of cleaner technologies and in the progress made in firms' EM.

As suggested in the received literature, the adoption of PP measures has economic advantages with respect to more traditional control methods and in many cases it has also resulted in net economic benefits, even though they are generally less profitable than investments made in non-environmental areas. The importance of local innovative and quality capabilities is also shown here – most PP projects have been developed by firms with such assets. This type of enterprise has also been responsible for the bulk of the more profitable PP projects. However, most of these projects have been based on 'simple' measures, in which productive and environmental efficiency seem to be synonymous. Finally, the costs and/or investment needs in EOP treatments have decreased as a consequence of the adopted PP measures but, except in a few cases, the use of EOP facilities have not been totally eliminated.

Summing up, the progress made in the 1990s in the adoption of more advanced environmental practices within Argentine industry seems undeniable. However, such progress has so far been concentrated in a small group of firms, especially large, export-oriented firms or TNC subsidiaries. Most Argentine manufacturing firms have made little progress in this field. Even among those firms that seem to have made greater efforts, the adopted measures are mainly the 'simpler' ones. Since good EM would increasingly become a crucial condition to compete in the market place, this situation is serious not only for environmental reasons but also for economic ones. This perception has not yet been incorporated among local businessmen, particularly among SMEs.

Moreover, even among large, export oriented and/or foreign firms, the advances made do not cover the whole spectrum of EM. Whereas in several cases accumulated problems have to be solved, in other cases measures involving higher costs and/or investments have to be implemented. Likewise, customer and/or supplier interactions need to be expanded for a greater and quicker diffusion of an EM based on product life-cycle criteria.

In this connection several issues should receive priority on the research agenda. First, a deeper analysis of the economics of adopting PP projects would be useful in order to learn more about their characteristics, development conditions and impacts. It is important to find out if, in addition to those projects related to water, energy and input savings, other PP projects implying higher investments and technological capabilities can also be profitable. Second, access to technology as a key obstacle to the adoption of PP measures should also be investigated. It would be important to learn whether the main constraints are due to supply restrictions, lack of information/resources or cost problems.

Regarding SMEs, it would be useful to learn the extent to which the diagnosis made above may be generalised to firms operating in other regions of the country. Likewise, the specific causes of EM weakness among this type of firm as well as the main obstacles to improving such management should receive more attention. Furthermore, the relationship between EM, adoption of PP measures and innovation capabilities as well as the connection among the first two and the organisational management schemes in each firm should be studied in more detail.

Taking into account the growing importance of TNCs in the Argentine economy, their EM should be studied in greater detail. In this connection, their 'environmental' co-operation schemes with other agents in the local economy should be assessed.

Regarding the policy implications, the need for a clear environmental regulatory framework and appropriate environmental enforcement mechanisms is beyond any doubt. There is also a need to develop educational and social awareness mechanisms with the aim of stimulating both customers and producers to appreciate the benefits of a less polluted environment. However, the research findings suggest that it is likely that these actions will be insufficient to ensure a sustainable development path. The improvement and preservation of the environment should become an integral part of a growth strategy. Hence, instead of confronting productive activities, environmental policies should be gradually designed and implemented by introducing mechanisms to facilitate a negotiating process between the state, private firms, non-governmental organisations and other social groups. Likewise, the regulatory framework should provide incentives to induce the firms to improve their environmental performance through innovative activities.

In this connection, current environmental regulations, which have been defined mostly in isolation, should be integrated with other policies in related areas. For example, it is pretty clear that firms' EM strongly depend on their innovativeness and quality management. Hence, improvements in innovatory and quality capabilities should be considered as a priority within any global environmental policy. At the same time, public policies, aimed at innovation and quality, should explicitly take into account the environmental dimension.

The search and adoption of cost-effective technologies (following a preventive approach) should become a priority in environmental policy; for example, OECD guidelines may be followed to examine which current policies constrain or facilitate this kind of approach. Moreover, since firms and sectors do differ in their EM, performance and problems, the policies in question should clearly differentiate between the appropriate standards and actions for each case. In particular, SMEs' environmental problems should become a priority for policy makers.

Macroeconomic policies also have a key role to play. Growing productive investments are crucial in stimulating firms to update their machinery and equipment incorporating at the same time cleaner technologies. In addition, capital market failures cannot be overlooked when considering the financial difficulties faced by SMEs and generally in the allocation of funds for some risky activities, such as the development of new technologies.

Finally, there is practically no public information on the environmental performance of firms and sectors. Since the availability of such information is a basic requirement in order to examine the weak and strong points in EM within the manufacturing sector, environmental authorities should make serious efforts to bridge the current information gap. The current Argentine environmental situation is not only a serious danger for human health and biodiversity preservation, but the behaviour of the economy is also affected by such a situation. Environmental authorities and private sector attitudes towards the environment need to be modified to change this state of affairs. Otherwise, the unavoidable adjustment would be more drastic and would mean higher economic and social costs.

Notes

1 The financial support of the International Development Research Centre of Canada, the North South Center of the University of Miami and the Avina Foundation and the collaboration of the Argentine Business Council on Sustainable Development (CEADS) to the research project on which this paper is based are gratefully acknowledged.

2 The first is a survey of 32 large firms operating in Argentina, prepared and distributed jointly with the CEADS. The second is a survey of 120 industrial SMEs in Gran Buenos Aires, carried out with the collaboration of Quilmes and General Sarmiento National Universities. The aim of the survey was to obtain a diagnosis of the evolution, current situation and perspectives of such firms in the present national economic context. Among the survey's different issues, some questions related to EM were included. However, in view of the main objective of the fieldwork and the kind of firms involved, EM issues received far less attention than in the survey to large firms.

3 See López (1996) for a comprehensive discussion.

4 Product offsets occur when environmental regulation produces not just less pollution, but also creates better-performing or higher-quality products, safer products, lower product costs, products with higher resale or scrap value or lower costs of product disposal for users. Process offsets occur when environmental regulation not only leads to reduced pollution, but also results in higher resource productivity such as higher process yields, less downtime through more careful monitoring and maintenance, material savings, better utilisation of by-products, lower energy consumption, reduced material storage and handling costs, etc.

5 Pollution-control technologies – which transform dangerous substances into harmless ones before being emitted to the environment – and clean-up technologies, which make innocuous those dangerous substances that have already penetrated into the environment and/or improve degraded ecosystems, are included in the EOP measures classification (NSTC, 1994).

6 However, in many cases, the dividing line between PP and EOP is imprecise; for example, sometimes substances of positive economic value are recovered by adopting EOP treatments. In addition, EOP treatment is seldom totally eliminated by the adoption of PP measures.

7 For example, a study on a number of TNC subsidiaries in the USA concluded that there was no correlation between environmental practices and pollutant emissions in each firm (Levy, 1995).

8 These figures are in line with previous information on Argentina and on other countries of the region. In Argentina, it had been estimated that, between 1990 and 1995, in high pollution potential sectors such as steel, petrochemical and pulp and paper industries nearly 10 per cent of total sectoral investment was related to environmental issues (Chudnovsky *et al.*, 1996). In Chile, at the beginning of the 1990s, approximately 6 per cent of the investment to build three plants in the pulp and paper sector was associated with environmental aspects (Scholz *et al.*, 1994). Motta Veiga *et al.* (1994) quote investments of US$ 400 million in the Brazilian steel industry between 1988 and 1992, implying a share of 7–10 per cent in the total investment of the branch.

9 However, according to Domínguez-Villalobos (1999), there are some exceptions, suggesting that the relationship between size and environmental behaviour is not absolutely lineal.

10 Although Brugger *et al.* (1996) compiled a set of cases about SMEs in Bolivia, Colombia and Costa Rica where there has been progress related to EM, broadly speaking, SMEs of the region have significant problems in improving their environmental performance.

11 It is important to note that most firms assigning greater importance to water, energy and input savings measures are included in medium/high or high-energy intensive sectors. The definition of sectors according to their energy intensity is based on Bezchinsky *et al.* (1992) who consider the relationship between energy consumption expenditures and gross production value.

12 To analyse and compare the firms' situation regarding the main research issues, four indicators have been defined to reflect, respectively, the EM level – both for large firms and SMEs, the adoption of PP practices and the firms' quality and innovation capabilities.

For large firms, the following variables have been included in the EM indicator: existence of a formal EM department; number of persons involved in environmental protection activities; monitoring of environmental indicators and goals; environmental targets; environmental accounting system; environmental investments registration; percentage of R&D expenditures geared to environmental issues; the implementation of studies on the environmental impacts of product/process and on raw materials recycling; analysis of possibilities of using environmentally friendly raw materials and/or technologies; interactions with customers and/or suppliers; adoption of PP measures; and environmental certifications.

For SMEs, the EM indicator has been built considering the following variables: knowledge of environmental regulations, targets for environmental performance and measures implemented to reduce pollution levels (such as EOP treatment, maintenance and operating practices, staff training, customer/supplier co-operation schemes, energy, water and input savings, product and/or process reformulation, raw material substitution, adoption of new clean technologies and external recycling).

The proposed PP indicator has been estimated through the importance given by the firms to the following measures: maintenance and operating practices, staff training, customer/supplier co-operating schemes, energy, water and input savings, product and/or process reformulation, raw material substitution, modifications of existing processes, adoption of clean technologies and external recycling.

Finally, the quality and innovative capabilities' indicator has been estimated through the following variables: R&D expenditures as percentage of sales, number of engineers, professionals and scientists as percentage of total employment and quality certifications either obtained or in progress.

13 Specialized foreign engineering firms are also an important source of environmental technologies for Chilean pulp producers (Scholz *et al.*, 1994).

14 According to Scholz *et al.* (1994):

> strategies to improve the competitiveness and productivity of the firm may contribute to strengthen its environmental adjustment capabilities. The complementary nature of these goals come from the fact that the aptitude to

> carry out techno-organizational processes of innovation and learning is an important premise to improve both the competitiveness and, at the same time, the environmental management capability of the firm.

Likewise, it was pointed out that "firms operating with higher levels of efficiency and having more innovatory capabilities can improve both environmental quality and productivity simultaneously. Moreover, environmental efforts could become substantial economic benefits" (ECLA, 1995). Similar conclusions can be found in the Argentine (Chudnovsky *et al.*, 1996) and Mexican (Domínguez-Villalobos, 1995, 1999) cases.

15 The classification of branches according to their technological intensity is based on the OECD methodology. Broadly speaking, this methodology takes into account, among other elements, the significance of R&D expenditures of total output.

Bibliography

Bezchinsky, G., R. Bisang and F. Eggers (1992) *Tabla de Categorización Tecno-Económica de las Producciones Industriales Argentinas.* Buenos Aires: ECLA (mimeo).

Brugger, A., A. Pinto and C. Barragán Castaño (1996) *Ecoeficiencia en la Pequeña Empresa, motor del desarrollo sostenible latinoamericano.* Colombia: PROPEL.

Chudnovsky, D. and M. Chidiak (1996) 'Competitividad y medio ambiente. Claros y oscuros en la industria Argentina', *Boletín Informativo Techint*, No. 286, April–June.

Chudnovsky, D., F. Porta, A. López and M. Chidiak (1996) *Los Límites de la Apertura. Liberalización, Reestructuración Productiva y Medio Ambiente.* Buenos Aires: Alianza/CENIT.

Dahlman, C. J., B. Ross-Larson and L. E. Westphal (1987) 'Managing Technological Development: Lessons from the Newly Industrializing Countries', *World Development*, Vol.15, No. 6.

Domínguez-Villalobos, L. (1995) *Reconversión Industrial y Medio Ambiente (estudios de casos de las firmas Dupont, VLX, Carp y Cydsa)*, prepared for the Seminar: 'Instrumentos económicos para un comportamiento empresarial favorable al ambiente'. México, DF: El Colegio de México.

Domínguez-Villalobos, L. (1999) 'Comportamiento Empresarial hacia el Medio Ambiente: El Caso de la Industria Manufacturera en la ZMCM', in A. Mercado García (ed.) *Instrumentos Económicos para un Comportamiento Empresarial Favorable al Ambiente de México.* México D.F.: Fondo de Cultura Económico/El Colegio de México.

ECLA (1995) 'Medio Ambiente y Comercio Internacional en América Latina y el Caribe', in SELA/UNCTAD, *Comercio y medio ambiente. El debate internacional.* Caracas: Nueva Sociedad.

FIEL (1996) *Medio Ambiente en la Argentina. Prioridades y Regulaciones.* Buenos Aires: Fundación de Investigaciones Latinoamericanas.

Hanrahan, D. (1995) *Putting Cleaner Production to Work.* Washington, DC: World Bank, Discussion Draft.

Katz, J. (1990) 'Las Innovaciones Tecnológicas Internas y la Ventaja Comparativa Dinámica', in S. Teitel and L. Westphal (eds) *Cambio Tecnológico y Desarrollo Industrial*. México, DF: Fondo de Cultura Económica.

Lall, S. (1992) 'Technological Capabilities and Industrialization', *World Development*, Vol. 20, No. 2.

Levy, D. L. (1995) 'The Environmental Practices and Performance of Transnational Corporations', *Transnational Corporations*, Vol. 4, No. 1.

López, A. (1996) *Competitividad, Innovación y Desarrollo Sustentable*. Buenos Aires: CENIT, DT No. 22.

Motta Veiga, P., M. Reis Castilho and G. Ferraz Filho (1994) *Relationships between Trade and the Environment. The Brazilian Case*. Rio de Janeiro: FUNCEX (mimeo).

National Science and Technology Council (NSTC) (1994) *Technology for a Sustainable Future*. Washington, DC: Office of Science and Technology Policy.

O'Connor, D. and D. Turnham (1992) *Managing the Environment in Developing Countries*. Paris: OECD Development Centre, Policy Brief No. 2.

OTA (1994) *Industry, Technology and the Environment: Competitive Challenges and Business Opportunities*. Washington, DC: Congress of the United States, Office of Technology Assessment.

Porter, M. and C. van der Linde (1995a) 'Toward a New Conception of the Environment – Competitiveness Relationship', *Journal of Economic Perspectives*, Vol. 9, No. 4.

Porter, M. and C. van der Linde (1995b) 'Green and Competitive', *Harvard Business Review*, Sep.–Oct.

Scholz, I., K. Block, K. Feil, M. Krause, K. Nakonz and C. Oberle (1994) *Medio Ambiente y Competitividad: el Caso del Sector Exportador Chileno*. Berlin: Instituto Alemán de Desarrollo, Estudios e Informes 13/1994.

UNCTAD (1993) *Environmental Management in Transnational Corporations*. New York: Program on Transnational Corporations, Environment Series No. 4.

World Bank (1995) *Argentina. Managing Environmental Pollution: Issues and Options*. Washington, DC: Environmental and Urban Development Division.

World Commission on Environment and Development (WCED) (1987) *Our Common Future*. Oxford: Oxford University Press.

5

'AÇO VERDE': THE BRAZILIAN STEEL INDUSTRY AND ENVIRONMENTAL PERFORMANCE

Jonathan R. Barton

The development of Brazilian iron and steel production[1]

Brazil's 'economic miracle' of the 1950s and 1960s focused on the São Paulo–Belo Horizonte–Rio de Janeiro triangle in the southeast and their manufacturing industries, particularly the automobile, metallurgy, chemicals and pharmaceuticals sectors. Amongst these, the metallurgy sector has been identified as the most environmentally degrading activity, based on a potential impact matrix developed by da Gama Torres (1996: 49), and iron and steel production is the cornerstone of this sector. Production is located across the southeastern states (IBS, 1997) with the major centres in Minas Gerais (38.7 per cent), Rio de Janeiro (23.6 per cent), São Paulo (18.5 per cent) and Espírito Santo (14.2 per cent).

The Brazilian steel industry dates back to the nineteenth century, but the sector was developed during the industrial development programmes of Presidents Vargas and Kubitschek during the 1940s and 1950s. Until the 1990s, the steel sector was predominantly state run, becoming highly inefficient and technologically dated during the authoritarian period (1964–85), but privatisation under the Plano Real has led to modernisation, improved technologies and more efficient management. Early production was based on the iron ore deposits, energy availability and charcoal resources in Minas Gerais, but the industry expanded into other states from the 1940s to cater for local demand and new employment demands. The construction of the Companhia Siderúrgica Nacional (CSN) at Volta Redonda was the jewel in the crown of this expansion, establishing the largest steel production site in Latin America, and other major plants were constructed in the 1960s (Cosipa and Usiminas) and 1970s (CST and Açominas).

By 1964, production had exceeded national demand for the first time, and the sector was increasing productivity and diversifying its products (Dantes and Souza Santos, 1994). This development phase was overseen by the Steel Industry Consultative Council under its National Steel Plan from 1968, and was supported by the BNDES (Brazilian National Development Bank) and foreign capital and technology, until a holding company – Siderbras – was founded in 1973 to co-ordinate state steel activities. As a strategic sector, the industry was protected by the government, and output and employment absorption were promoted to the exclusion of other considerations, such as environmental impacts.

The most significant changes came in the early 1990s with the introduction of the Plano Real. Part of the new era of monetary control involved curbing public sector spending and privatising public sector firms. The steel sector was an early target of the Plan. Siderbras was dismantled and its operations sold off, beginning with Usiminas (1991), CST and Acesita (1992) and finally CSN, Cosipa and Açominas (1993). Most of the new investment was domestic and led to the revitalisation of the firms which had lacked capital during the 1980s (Pinheiro da Silva, 1992).

Currently, production is balanced between the larger integrated plants producing semi-finished steels principally for export (see Fig. 5.1), flat products principally for the auto industry, and long products for construction and infrastructure works, and the semi-integrated plants which concentrate more on long products; special steels such as stainless remain a small but increasing area of production for both types of plant (see Table 5.1).

Much of the new investment has been targeted at the flat products' market for the auto industry and special steels (mainly stainless) (see Fig. 5.2). Long products, destined principally for construction, and semi-finished products

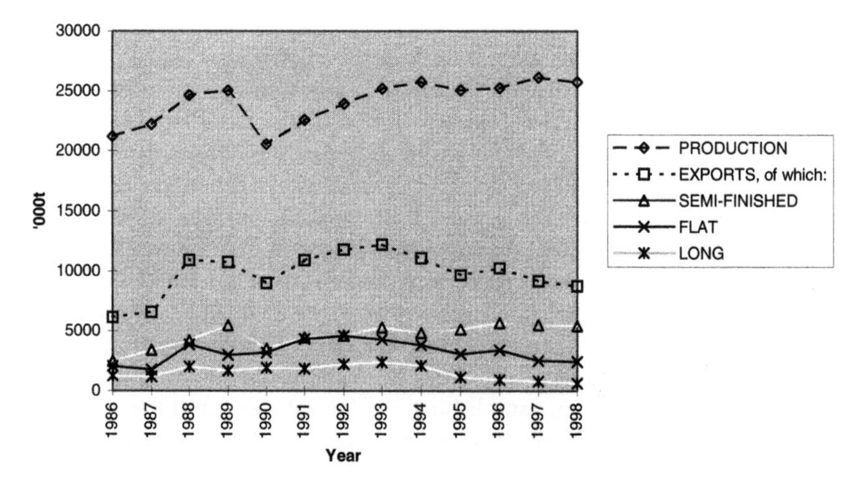

Figure 5.1 Brazilian steel production and exports, 1986–98. Source: IBS (various years) *Anuário Estatístico.*

Table 5.1 The Brazilian steel sector, 1998

Products	*Integrated plants – capacity 24.4Mt/yr (ironmaking and conversion to steel)*
Semi-finished (flat and long)	Açominas (MG), CST (ES)
Flat	Cosipa (SP), CSN (RJ), Usiminas (MG)
Long	Belgo-Mineira (MG), Gerdau (MG, BA)
Special steels (flat and long)	Acesita (MG), CST (ES)
	Semi-integrated plants – capacity 5.8Mt/yr (principally conversion of scrap into clean steel in electric arc furnaces, EAF)
Long	Gerdau (PR, RS, CE, PE, RJ), Belgo-Mineira (SP, ES), Barra Mansa (RJ), Mendes Jr./BMP (MG), C.B.Aço (SP), Itaunense (MG)
Special steels (long)	Aços Villares (SP), Villares Metals (SP), Gerdau (RS)

Source: IBS (1999).
States: SP – São Paulo; MG – Minas Gerais; ES – Espírito Santo; RJ – Rio de Janeiro; BA – Bahia; RS – Rio Grande do Sul; PE – Pernambuco; PR – Para; CE - Ceará.

are the other main product categories. This chapter focuses on the five largest steel firms in Brazil which produce flat products destined predominantly for the domestic market or semi-finished products predominantly for export. The two other groups (long-product integrated plant producers and electric arc furnace (EAF) firms) have significantly smaller output levels. For EAF producers, comparisons with integrated works are fruitless since the processes and pollution issues are wholly different. There may be differences between the

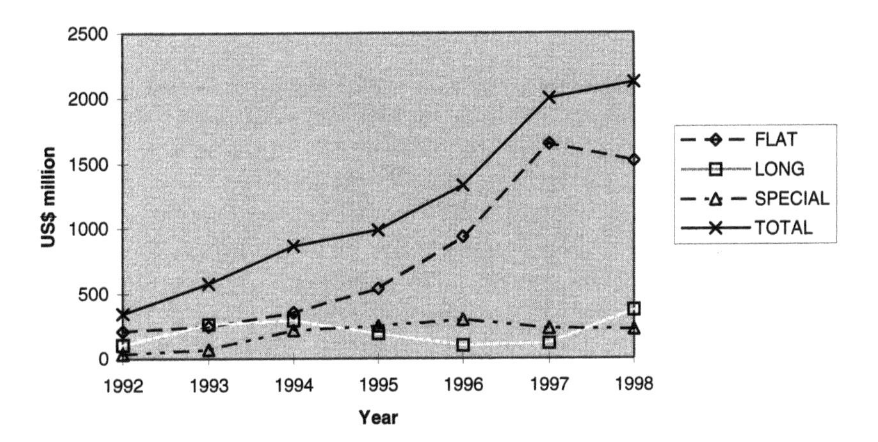

Figure 5.2 Investment expenditure by the steel industry. Source: IBS (1999).

long-product group of firms since they produce predominantly for domestic markets, but the research did not extend to these companies.

Prior to privatisation, the steel companies had been managed with political and strategic objectives and were beset with problems, such as production lines limited by insufficient inputs, supplier and transport cartelisation, poor labour relations, steel prices indirectly managed by the government via buyer subsidies, and a negative and discredited image with respect to supply times and quality. The privatisation process enabled the firms to establish a clearer commercial footing and move beyond these obstacles. This included product diversification (e.g. CSN's two-piece steel can), greater emphasis on client service, adding value to products (e.g. painting and tinning) and steel production optimisation (via quality controls and new technologies). Firms both 'verticalised' in terms of attention to their clients and in terms of moving upstream to control their suppliers by buying shares in other privatised companies which provided energy, transport and raw materials. For instance, CSN bought shares in energy and transport providers, a 41 per cent share of Companhia Vale do Rio Doce (CVRD) (within a consortium) – the world's largest iron ore supplier and Brazil's leading exporter, and a 25 per cent ownership of the Ribeirão Grande cement firm to profit from the conversion of their own slag by-products.

The learning curve has been steep, requiring labour reductions for productivity reasons (132,000 in 1990 to 78,000 in 1996 across the sector – IBS 1997), heavy new investment in equipment, tighter controls over costs, and new strategies for marketing, sales, and materials and product development. Alongside increasing their own competitiveness as firms, companies and the IBS have been pressurising the government to reduce the obstacles to competitiveness that are present within the Brazilian systems of finance, taxation, tariffs, working practices and infrastructure. These are commonly referred to as the 'Brazil Cost' (see de Oliveira and Tourinho, 1996) and are a hangover from the years of public ownership in infrastructural, energy and transport activities.

A positive factor for the steel industry is that new investments following privatisation have enabled companies to adopt a more aggressive stance relative to modernisation and development than in other steel producing countries where capacity is not increasing. The current low per capita steel consumption (compared internationally) allows considerable scope in this respect (see Fig. 5.3). CSN, for example, made an investment of $730 million in a new mini-mill in Pecém, Ceará, to take advantage of local raw materials and port facilities. Other companies have adopted different strategies, depending on their products and circumstances. CST, which produces semi-finished steel almost entirely for export, dominates a low price, high volume market by producing the lowest cost liquid steel in the world (CST, 1997). This low-cost export strategy has taken the company to third place among Brazil's export companies (behind CVRD and Ford, 1996).

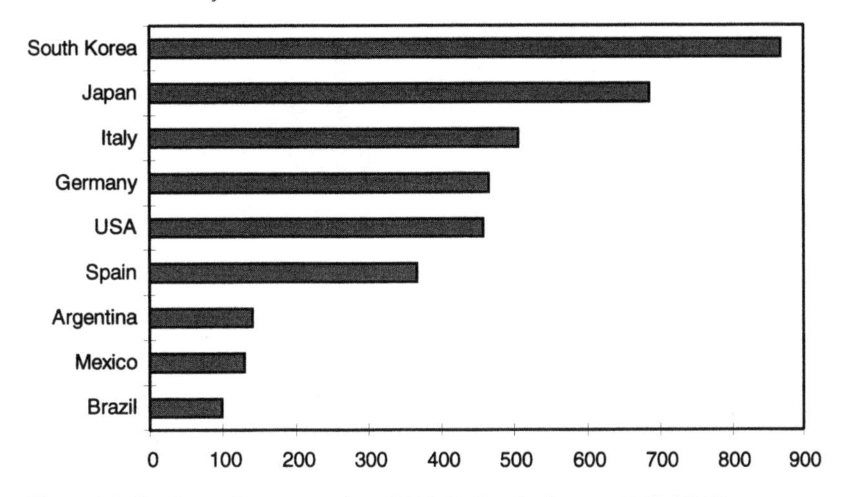

Figure 5.3 Crude steel consumption, 1997 (kg/cap). Source: IBS (1999).

Although production levels have remained relatively stable during the post-privatisation period, the levels of investment have been considerable and have led to the updating and modernisation of the existing plant and investment in technologies to add value to the product range; steel firms have also invested in energy provision and now provide 14 per cent of their demand, against a figure of only 10 per cent in 1990 (IBS, 2000). A new post-privatisation emphasis has also emerged in terms of environmental awareness. Most companies have increased environmental expenditure, even those which were already private and not part of Siderbras.

Environmental impacts and production factors

Iron and steel production is widely regarded as a 'pollution intensive' industry. Even disregarding raw materials, scrap and chemicals' storage issues, the central processes lead to a range of pollutants and involve a variety of pollution media. These are concentrated in the ironmaking phases of production, followed to a lesser extent by the conversion of iron into liquid steel and semi-finished products in the steelmaking process. Downstream rolling of these semi-finished products is comparatively less polluting and varies considerably in terms of the technologies and techniques employed, and the particular products (see Fig. 5.4; Tables 5.2 and 5.3). Emissions and discharges vary according to scale and process (integrated and EAF) and pose diverse problems for firm environment managers, some of which can be ameliorated via technological change and others which require improved process operation and a focus on better management, training and procedures.

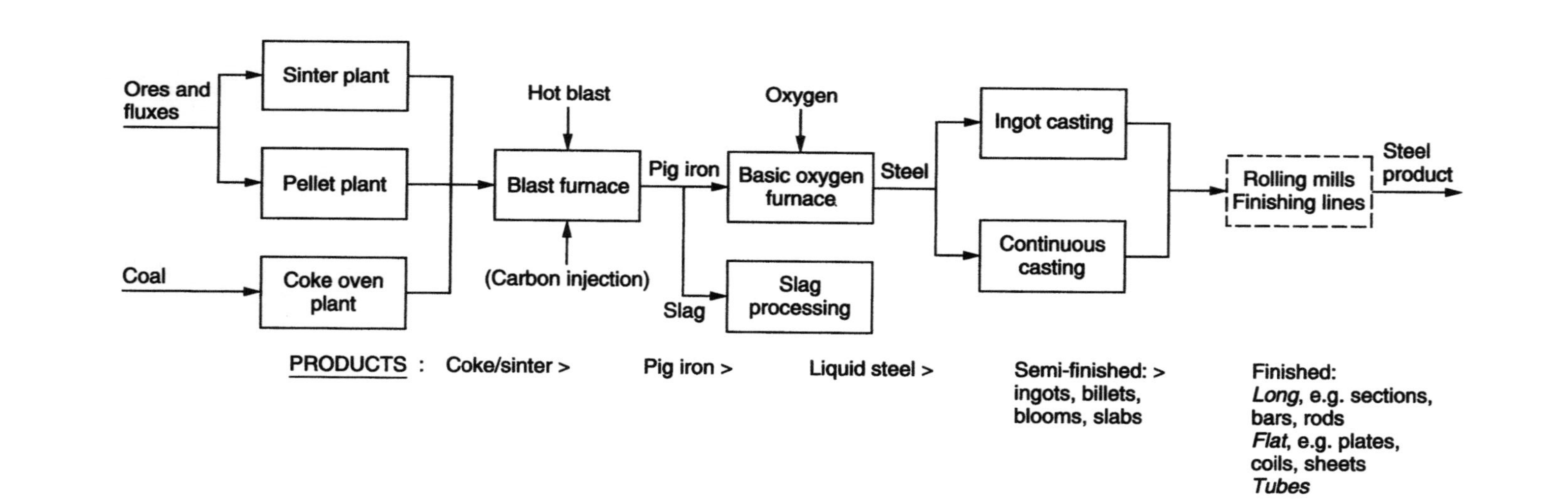

Figure 5.4 The primary iron and steel production process.

Table 5.2 Environmental impacts of steel manufacture (pollution-intensive processes)

Process stage	*Potential pollutant release*	*Potential environmental impact*
Sinter/pellet production	Dust (inc. PM_{10}), CO, CO_2, SO_2, NO_x, VOCs, methane, dioxins, metals, radioactive isotopes, HCI/HF, solid waste	Air and soil contamination, ground-level ozone, acid rain, global warming, noise
Coke production	Dust (inc. PM_{10}), PAHs, benzene, NO_x, VOCs, methane, dioxins, metals, radioactive isotopes, HCI/HF, solid waste	Air, soil and water contamination, acid rain, ground-level ozone, global warming, odour
Blast furnace	Dust (inc PM_{10}), H_2S, CO, CO_2, SO_2, NO_x, radioactive isotopes, cyanide, solid waste	Air, soil and water contamination, acid rain, ground-level ozone, global warming, odour
Basic oxygen furnace	Dust (inc PM_{10}), metals (e.g. zinc), CO, dioxins, VOCs, solid waste	Air, soil and water contamination, ground-level ozone
Electric arc furnace	Dust (inc PM_{10}), metals (e.g. zinc, lead, mercury), dioxins, solid waste	Air and soil contamination, noise
Waste water treatment	Suspended solids, metals, pH, oil, ammonia, solid waste	Water/groundwater and sediment contamination

Source: Ministry of Housing, Spatial Planning and the Environment, The Netherlands (1997).

Table 5.3 Integrated plant: principal emissions (%) – contributions by process

		Ironmaking (excl. coke plant)		*Steelmaking*	*Finishing*
Emission	*Total (g/t)*	*Sinter plant*	*Blast furnace*	*BOF and CC*[a]	*Rolling*
Dust	640	69.5	13.3	7.0	10.1
CO	27,280	92.7	1.3	5.9	–
SO_2	1830	66.9	7.9	–	25.1
NO_x	1050	60	8.5	0.9	30.5
Landfill	51 kg/t	–	9.8	81.4	8.8

Source: EC (1996).

Note

a BOF: basic oxygen furnace; CC: continuous casting.

Gauging precise environmental impacts of the steel industry and therefore changes in environmental performance is highly problematic. The problems revolve around monitoring and data collection which vary between companies, the geophysical conditions of the plant sites, production differences (in terms of raw material and energy mixes, process and output) and the demands of regulators and municipalities. For each company, investment in and management of these environmental problems is also conditional upon the firm strategy and its business cycle, regulatory demands, and market and local stakeholder pressures. These demands lead to different prioritisation relating to atmospheric emissions reductions, water discharge concentrations and solid waste recycling and sales.

In different pollution media, there are particular issues to be resolved. Atmospheric emissions present the principal problem due to CO_2, SO_x, NO_x, dust and heavy metals regulations. The approach to regulatory compliance has traditionally been technological, with the application of 'end-of-pipe' capture equipment such as bag filters and electrostatic precipitators, dependent on process stage, equipment age and the particular circumstances of emissions and local regulatory variations. Beyond emissions relating to combustion processes, further areas of control include particulate materials raised from stockyard raw materials and during transportation. This is especially important for those firms operating in the proximity of residential areas with complicating meteorological conditions.[2]

Water treatment was prioritised until the 1970s owing to the earlier regulatory attention to discharges, but most companies have since installed effective physical, chemical and biological wastewater treatment systems. Particularly contaminating effluents emerge from cokery operations, gas scrubbing water, and water from finishing activities. For the most part, water treatment is well advanced within the sector; however, strict targets for particular pollutants, especially ammonia, pose problems. In terms of water use, firms have improved the recirculation of water around the plants to lower demand and this is likely to intensify with new legislation charging for intake as well as discharge.

Air and water emissions control leads to cross-media impacts, in particular an increase in by-product and waste production and the need for recycling. The increases in landfill charging in the 1990s have fuelled research into and piloting of by-product reutilisation and treatment. Markets are sought for by-products such as blast furnace and steelmaking slags, but certain by-products, such as oils, are Class I category materials which are hazardous and incur heavy landfill costs. To assist firms in disposing of by-products more cheaply and effectively, the Brazilian Steel Association (IBS) has been pressurising the federal environment body, Conselho Nacional do Meio Ambiente (CONAMA) to reclassify certain by-products to facilitate recycling and sale.

Iron and steel production is energy and raw material (coal and iron ore) intensive; these inputs account for over 60 per cent of production costs (Booz-

Allen and Hamilton, 1996). Since the gains from improved output per unit of raw material are limited, it is energy that has become the focus for eco-efficiencies. All steel companies are proactive in reducing energy demand primarily for cost-based reasons; however, there are positive environmental outcomes from these changes in demand and energy mix. Brazil lags behind other producers in energy consumption per unit of steel; however, the increasing application of continuous casting and other production technologies will ensure a continuing downwards trend to compete with other international producers (Table 5.4).

Besides the energy issue, there is an element of steel production that is particular to the Brazilian case – the extensive use of charcoal for coking. The lack of suitable coal deposits in the country has led to large imports during the 1990s. Prior to privatisation, firms were obliged by Siderbras to use domestic coal, but this material was high in sulphur and inefficient. The environmental impacts of charcoal use in steel production relate to the native forest loss and replacement with homogeneous, exotic species, and also to the impacts of the charcoal production process, and transport of the material and its final use (Medeiros, 1995). Despite criticisms, many large firms have developed R&D programmes to establish sustainable inputs into their production systems (Mannesmann Florestal Ltda, 1996). These large steel firms have generated respectable sustainable charcoal operations with mainly eucalyptus species, and must be contrasted with many *guseiros* – small-scale pig iron producers, which consume up to three times more charcoal than steel firms (see Fig. 5.5). Despite sustainable forests however, steel industry consumption of charcoal has declined during the 1990s from 9.2 million m^3 in 1990 to 5.3 million m^3 in 1996 (IBS, 1997), and there is no evidence that they have adopted a strategy of purchasing rather than producing charcoal-based pig iron (see Fig. 5.6). With extensive investment in managed plantations, it is unlikely that charcoal will disappear from steel production in the medium term, but the principal issue is what constitutes a sustainable balance between exotic and native species and what regulatory limits on production should be imposed.

Table 5.4 Gigajoules per tonne of steel produced (average)

France	20.0
CST	20.1
USA	20.1
Germany	17.6
Japan	18.1
Brazil	24.9

Source: CST (n.d.) *Environmental Profile*.

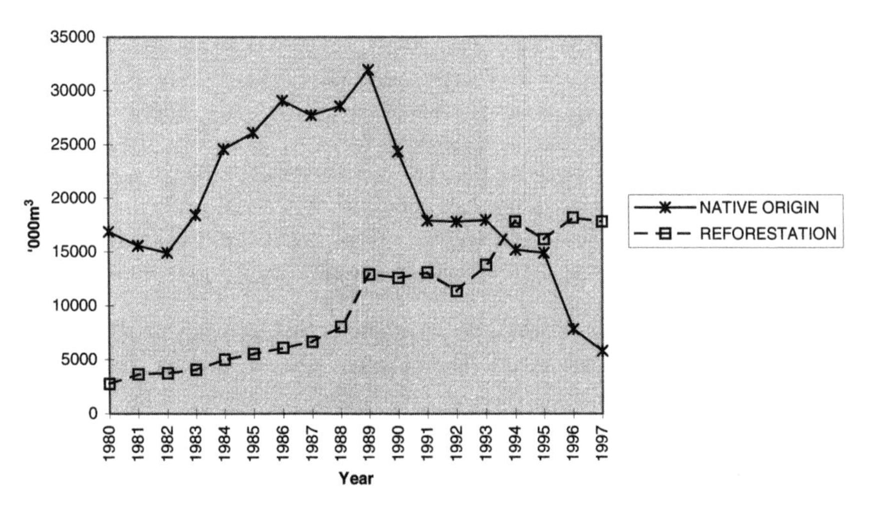

Figure 5.5 Charcoal supply in Brazil. Source: IBS (various years) *Anuário Estatístico*.

Once one includes upstream environmental impacts from charcoal production, coal mining and the mining of iron ore, limestone, dolomite and other materials used in steel production, the picture generated of the steel sector is different from a site-only perspective. Regulatory agencies lead firms to focus on site emissions and discharges of pollutants; however, there is increasing attention paid to 'life cycle inventories' in the steel industry, as conducted by the International Iron and Steel Institute (IISI), and these will reveal both the environmental impacts derived from production inputs as well as some of the eco-efficiencies derived from increased liquid steel output per unit of input

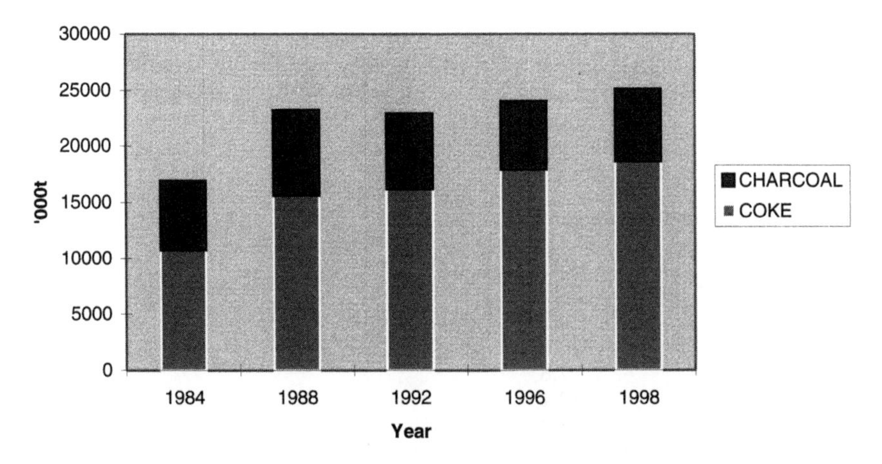

Figure 5.6 Pig iron production by process. Source: IBS (various years) *Anuário Estatístico*.

and the potential benefits from higher recyclability rates of steel products. With the industry concentrating on marketing steel products relative to its competitor products (plastics and aluminium), the ability to reveal a lower environmental impact per unit of production within a 'cradle to grave' framework will become increasingly important.

Environmental performance in the Brazilian steel industry

Environmental management requires a range of skills, technologies and programmes: procedures must be defined; management systems put in place; technology controls established; modernisation of existing equipment promoted; pollution control and operational efficiency optimised; accidents prevented and controlled; and green areas developed, monitored and maintained (IBS, 1990, 1995). It is a relatively new area of management expertise in Brazil and elsewhere following the technological approaches to environmental performance from the late 1970s. During the 1990s, more emphasis has been given to establishing departments and personnel to oversee implementation, monitoring (including audit and evaluation) and effective environmental public relations and education.

Ronaldo Campos Soares (1990), president of Usiminas, sums up the three pillars of environmental management as: integration with regulators, relations with the community and management within the company. With this strategy, Usiminas has been a leader in industrial environmental management and the firm stays ahead of regulators and competitors with clear objectives and action plans that go beyond technological applications towards reduction targets and practical environmental policy declarations. The ISO 14000 norms have become an important driver since firms equate certification with international market access opportunities (see Centro Brasileiro da Qualidade, Segurança e Productividade, 1994; Claudio, 1994; Cetesb, 1995); this explains why the firm Gerdau, which serves the domestic market, has not prioritised certification. Other 'advantages and opportunities' associated with certification include (Usiminas, 1997, n.d.):

- rationalisation of human, physical and financial resources
- competitive advantage of preservation and expansion of markets
- reduction of costs and increased productivity
- reduction of insurance premiums
- revitalisation of the Quality Guarantee System
- legal security
- proactive posture regarding regulators and community
- access to lines of credit
- prevention and control of accidents and losses
- increased environmental consciousness
- market advantage via company image

Although Usiminas is the only steel company to be accredited (as of December 1997 the culmination of US$350 million of environmental protection investment),[3] others are going through the initial steps towards accreditation, such as auditing and the construction of systems and procedures. Environmental management systems vary although they have core activities that are common (Table 5.5).

Often the construction of an environmental management system (EMS) leads to more positive approaches to environmental management and assessments of the costs and benefits of major environmental investments. Since its founding in the early 1940s, CSN has invested US$200 million in environmental prevention and control, divided between US$105 million – air, US$75 million – water, and the remainder in landfilling and noise control. Of this figure, US$60 million was expected to be invested between 1994 and 1999 (CSN, 1997). CST invested US$231 million during construction (1983), US$119 million as ongoing capital expenditure and US$55 million in new investments, while Cosipa (1997) has invested in the environment in the form of its PAC (Projectos Ambientais Cosipa) fund, set up from the beginning of 1997 and which has invested US$160 million in pollution control (23 per cent of a total investment plan of US$691 million).[4]

These costs can bring about swift environmental improvements, especially within the Brazilian context of low environmental investment during the 1980s (despite an accumulated rise from US$400 million in 1980 to US$1340 million in 1990) (IBS, 1991). For example, Almeida e Silva (1995) notes the benefits accruing from the CST's environmental policy: a 90 per cent atmospheric emissions reduction (1990–4); US$26 million saved by recycling and consequent reduced waste volumes; water recirculation of 91 per cent; and reuse of gases for 90 per cent of electrical energy production. These targets have been met from changes initiated by management directives led by the Business Council for Sustainable Development. Following their guidelines, the company has policies both within the company (com-

Table 5.5 Breakdown of the EMS of CSN

EMS	Written procedures integrated with productive areas
Monitoring	To ensure compliance with existing legislation
Policies	Emphasising the company's objectives relating to the environment
Institutional strategies	Actions connected with government organisations in order to influence situations which affect the company
Priority areas	Particulate stack emissions, particulates in suspension in air around the plant and city, measurement of effluents (especially pH, temperature, cyanide, phenols, ammonia, BOD, heavy metals), water treatment and quality stations for the Paraíba do Sul river

Source: *Metalurgia & Materiais* (July 1997).

pliance, auditing, EMS improvements, education) and external to the company (community dialogue, information dissemination, R&D co-operation). The Gerdau Group (1996) has invested US$46 million on 'eco-efficiency' (industrial water treatment, noise reduction and dust collection systems). The large privatised firms have invested in technology and environmental management, particularly ISO 14001 certification (achieved by Usiminas) and its implementation phases. Technologies take the lion's share of the capital: CST has included environmental protection in all three of its post-privatisation stages (1993–2000 total investment: $1.7–1.8 billion) (CST, 1997); and CSN has included environmental protection (6 per cent) as part of its US$1.3 billion technical development strategy (1995–2000; CSN, 1997). These environmental developments have been responses to more demanding regulatory activities, 'stakeholder' pressures (from workers and neighbouring communities) and pressures from the marketplace and wider civil society.

In order to implement environmental investments and EMSs and gain returns from them, there has been an intensification of personnel reorientation, within the organisation of environment teams and departments, and also in terms of training. Monitoring, maintenance and communication activities are the basic functions of environment departments although these vary in scale from fully fledged departments to smaller environmental sub-units, also the decentralisation of environmental responsibilities to facility level. Many companies have linked environmental issues with other areas such as health and safety, e.g. Cosipa in its Superintendency for Medicine, Worker Safety and Environment, while others seek to integrate environment and production managers and operatives, i.e. Usiminas has environment personnel in its Management for Environment and Urbanism, in Production Management, in an Environment Group and in the Laboratory for Utilities and Environmental Control. The training of workers and management in environmental matters has only been incorporated into the training programmes of firms during the mid-1990s, although environment managers appear to be convinced that once awareness is raised, energy savings and greater environmental risk awareness will be both economically beneficial and reduce accidents. In all these activities, EMSs, departmental support and training, and the role of executives is critical. In the case of Usiminas, the proactivity of the company president was the key to certification and the firm-wide environment strategy. The absence of detailed costings of environmental management and technology investments and their effectiveness may well explain the lack of interest shown by most executives.

The concept of a 'pollution cost system' (Maciel, 1993) is still underdeveloped in the steel sector. The consideration of pollution costs as a factor in production cost reduction and greater efficiency is gaining ground slowly although the difficulties associated with identifying and calculating specifically environmental costs, apart from production, energy and labour costs, are problematic. Items to be considered include by-products' recycling, sales and landfill, inputs (e.g. water) and emissions charging, compliance failure penalties, equipment maintenance and depreciation. Only with these types of calculations will it be possible to verify arguments, such as those of Porter and van der Linde (1995), that there may be a 'win–win' situation in terms of market advantages and overall benefits resulting from environmental protection investments and 'first mover' activities.[5] Related to accounting is the need for effective auditing, and most companies have been moving forward in this direction as part of the EMS with external consultancy guidance. CST's auditing was a response to state legislation in Espírito Santo from August 1996 which required three audits in three years and the publication of the audit in a leading newspaper as an 'Environmental Declaration' (Freitas Alvim and Morimoto, 1995).

Management issues are increasingly important within company environment strategy, but the technological approach remains central owing to the high fixed capital costs involved. Those firms which are more recent and were constructed with environmental technologies in place have gained considerable advantages over those older facilities which require retrofitting and upgrading. The costs entailed in the environmental improvement of older process technologies are considerable due to issues of space availability, physical interference and facility shutdown to allow installation (IBS, 1990, 1995). This has been a problem for many Brazilian firms.

Technologies can be separated into process and production methods (PPMs) and 'end-of-pipe' capture systems; however, this division often obscures the changes in environmental performance across a site since only explicitly environmental technology investments are calculated as 'environmental' while practically all process technologies lead to energy savings and lower emissions, but are not calculated in the same way. The outcome is a distortion of the relationship between total investment in environmental technologies and the overall trends in emissions' reductions. Encouraging firms to concentrate on inputs and cleaner production will further complicate this calculation, but will certainly bring longer-term environmental performance improvements. What is required is a more flexible evaluation of technology costs and depreciation, and their environmental impacts. The example of continuous casting of steel is a case in point. Rather than the traditional cooling of semi-finished products which would then be reheated for downstream rolling, steel is cast into the desired shapes as hot metal which removes the energy requirements and emissions' implications of the cooling and reheating phases. Another example is the installation of PCI (pulverised carbon injection) into blast furnaces which increases the efficiency of the combustion process within the furnace.

Atmospheric emissions' control technologies have demanded the highest investments for firms in their pursuit of lowering greenhouse gases and dust targets. The application of capture systems, predominantly bag filters and electrostatic precipitators, has led to dramatic reductions from a base year in the late 1970s when the energy price 'hikes' heightened consumption awareness. However, there are diminishing returns from these investments over time as the principal emissions stacks are fitted with the equipment and regulatory limits demand further reductions. CSN, for example, has invested US$14 million in dust emissions' reductions by installing sinterisation-plant electrostatic precipitators (an increase of 6 tonnes/day of dust captured). Once all sinter stacks are fitted with the precipitators, the dust problems from 'fugitive' or secondary emissions around the equipment and from stockyards have to be confronted. To achieve reductions in these areas is also costly with lower returns per unit of investment. As 'end-of-pipe' technologies become universally applied, the emphasis then shifts from application to maintenance and good management. It is in this area of maintenance that many of the variations between firms can be found since similar technologies can be operated in different ways and are efficient only within certain limits and if well maintained. There has been an evolving recognition more generally that effective environmental management puts emphasis on process technology management, input selection and human resource management rather than purely on capture systems. However, investments in capture systems are likely to remain high until innovative strategies can be developed and are supported by top management and proven in terms of cost–benefit.

In the same way that primary stack emissions' technologies are widely employed for atmospheric emissions but vary in performance according to management systems, wastewater treatment can also be variable. Much of the concern for firms in terms of water discharge controls emanates from the plant location. For firms operating in inland areas within river catchments, such as CSN and Usiminas, the water discharges require more control than is the case with coastal sites. In the case of CST, for example, the company is located next to the sea and 95 per cent of its water requirements are met from that source; of the remainder, 91 per cent of the freshwater used is recirculated (CST, 1997). The pressures on inland sites are likely to intensify with new environmental policies. In Rio de Janeiro state, the regulatory agency, Fundação Estadual de Engenharia do Meio Ambiente (FEEMA) is developing a strategy based on a catchment. For CSN, as the principal user and discharger in the Paraíba do Sul river basin, this will imply greater coordination with regulators and other municipal and industrial user groups.

Increasingly, waste management is rising up the industrial environment agenda due to the costs of treatment and the disposal of large volumes of byproducts (according to toxicity). The industry has responded proactively, seeking buyers and ways of reutilising materials within the furnaces. The sector produces 700 kg of waste per tonne of steel, and recycling and commercialisation runs at 62 per cent (Maumédio de Paulo, 1992). Some companies, for various reasons, have been able to improve upon these figures (see Table 5.6).

Table 5.6 Solid-waste management at Usiminas and CST

Management of residuals	*Usiminas (%)*	*CST (%)*
Recycled, e.g. unprocessed in sinter plant, such as blast furnace dust and wastes, steelmaking dust and wastes, raw material fines	48	39
Commercialised, e.g. blast furnace slag to ceramic and cement industries, limestone for agricultural use	40	52
Controlled landfill disposal, e.g. Class I – oils; Class II/III – refractories, steelmaking, slags, acid neutralised waste, blast furnace slag, industrial waste	12	9

Sources: Usiminas (1996); CST (1997).

Apart from emissions and by-product controls, the energy issue has concerned firms since the 1970s. The two areas in which this is taking place are the supply of energy to the production process and in the improved efficiency and recycling of energy once in the process (see Fig. 5.7). Several companies are seeking to extend or rebuild their thermoelectric plants for reasons of modernisation, emissions limits and planned capacity increases. In the case of CSN, the development of a 235 MW thermoelectric plant to take advantage of gases generated in the steel process is an important step towards recycling energy at the plant level. Other elements of their energy programme revolve around participation in hydroelectric developments (at Itá

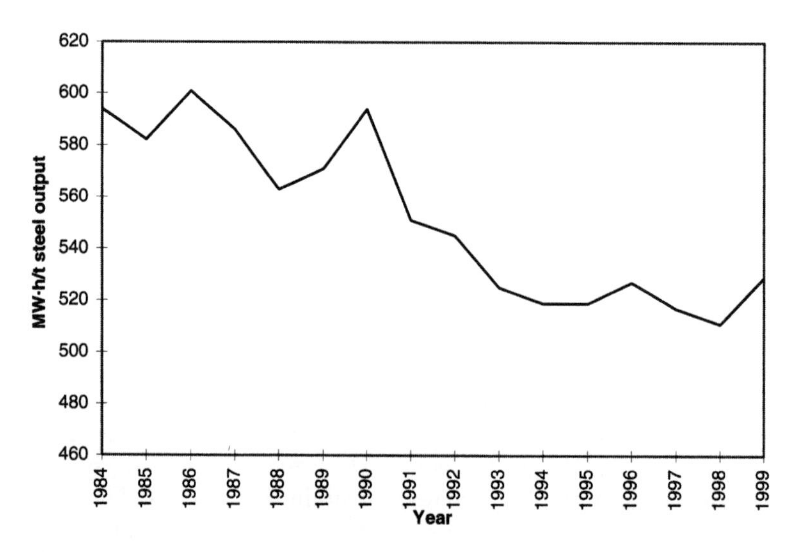

Figure 5.7 Electric energy consumption per tonne of steel, 1984–99. Source: IBS (various years) *Anuário Estatístico*.

and Igarapava). CST has focused on gas (coke oven gas and blast furnace gas) reutilisation, alongside producing 100 per cent of its own energy (150 MW) which will be increased by 120 MW with a new power plant (CST, 1997). Other companies are also seeking greater energy security alongside reducing consumption via recycling technologies.

The range of technological and management innovations has been extensive, and firms have required competent environmental management teams working across a range of tasks, including EMS, auditing, training, equipment maintenance prioritisation, establishment of 'environment' in the area of operational control, anticipation of regulatory visit problems, improved environment–operational personnel communications, and full legislative compliance. As legislation, consumers and pressure groups become steadily more demanding, the role of environmental protection and better cost–benefit analysis of management choices relative to this protection will become more focused.

Beyond regulatory compliance issues (public policy) and factors internal to the firm such as process and production method changes (production policy), there is a strand of environmental management that is influenced by the market and social considerations (marketing and public relations policy). This strand includes the role of the market, such as market access and product development, and also what may be termed as 'greening' and community programmes. The IBS (1990: 25) puts it in the following terms:

> The company must emphasise that it is conscious of its social responsibility and is preoccupied with promoting and harmonising the aspects linked with this responsibility, such as development and environmental protection.

The idea of 'greening' production sites and creating a green corridor between industries and residential communities was well developed during the 1960s and 1970s and has been a feature of all recent steel plant constructions and planting programmes for older sites. Trees and shrubs not only perform an aesthetic function but also reduce noise and dust transfers beyond the site. Since it is these two factors that are the most influential in terms of community opposition to steel firms, 'greening' programmes are important strategies. Under state ownership, firms were unlikely to respond to community or municipality pressures relating to environmental protection, but this has changed extensively since the end of authoritarianism. There is now recognition that communities are part of a network of so-called 'stakeholders' which firms have to pay heed to for reasons of public relations, firm image and ultimately wider financial and market reasons. Other stakeholder groups include steel users, workers and shareholders.

The need for industry–community environmental links emerged during the late 1980s and the IBS concedes that often the positive effects of industries are national while the negative effects can be concentrated locally, whether as a result of normal industrial operation or emergency incidents. For this reason, communication channels are essential. This may be especially important for large firms operating alongside large urban areas where traditional worker–community links have broken down as a result of rationalisation (IBS, 1990, 1995). Mechanisms for increasing industry–community links are numerous, including local newspaper information, company–residents' meetings, schools' visits and competitions, tree planting and, more urgently, 'hotline' telephone numbers in the case of emergencies or irregular pollution events. In many ways, the mechanisms seek to promote positive elements of environmental change with information dissemination, in order to compete with the negative sentiments that are circulated within the community and by NGOs and critical media.[6] The extent to which these mechanisms actually deal with the key environmental performance issues or are merely 'green window-dressing' to appease potential sources of opposition is debatable and probably varies with each case.

Almost all firms have some type of community-oriented environmental programme in place. Usiminas has one of the longest established environmental education programmes in the country. Its Projeto Xerimbabo is an adult and children's environmental education programme linked with the company's own Zoobotanic Park and has been running for 12 years. CSN and Cosipa are sited in areas of higher industrial concentration and have different programmes, such as CSN's monitoring strategy for the city of Volta Redonda and its extensive greening policy (currently 443,000 m^3 – CSN, 1997). Cosipa has also developed its green area policy and has achieved 17 m^2 per employee (the WHO recommends 12 m^2 per employee). More strategically important for Cosipa has been the Winter Operation 'Road Bath' scheme with various other agencies, which aims to prevent resuspension of dust in the Vale do Mogi area (Cosipa, 1997).

In terms of community linkages, there is also a tendency towards cultivating government–university–company networks. In the case of CST, there has been collaboration with the Federal University of Espírito Santo in the area of industrial development (see Morandi, 1995) and environmental control (see Guimarães, 1996). The establishment of the COEG (Organising Commission of Government Institutions) to link together municipal organisations and the state regulator in Vitória in order to oversee the fulfilment of the *Termos de Compromisso* of CST and CVRD (responsible for 90 per cent of the atmospheric contamination of Greater Vitória) is an example of this. Universities and research institutes have also been conducting important work on environmental impacts, recycling and environmental technologies, as well as instructing the next generation of environmental engineers and managers.

Factors influencing environmental performance

Regulatory control

Pressures for improving industrial environmental performance have emerged from two sources – regulations and social influences (people as neighbours and consumers). Although companies are keen to stress their environmental improvements and increased awareness (via reporting, for instance), there are still areas of concern in most plants in terms of process controls and emissions' reductions. A problem for analysts of industrial environmental protection is the rhetoric of company documentation and claims, and the realities of emissions' levels which have to be considered in detail (and are not readily available) to determine performance. In view of their access to firm information, and as the conduits of public pressures for 'cleaner' industries, it is the regulatory agencies which are best placed to make these judgements and to pressurise firms to manage environmental problems more effectively, both in a supportive way and with a 'big stick' if necessary.

The government agencies in charge of environmental regulations and their implementation vary according to states. Resources, styles of monitoring and enforcement, and environmental performance strategies also differ across boundaries with the result that production sites with similar environmental impacts are treated differently. The separation of federal and state regulations has played an important role in this situation since federal regulations provide a minimum framework for environmental performance and individual states have the legal authority to enforce stricter requirements. The outcome is that the location of industrial sites is becoming increasingly responsive to environmental regulations and their enforcement in different states (see Neder, 1996). Since competitiveness and production costs now include the actual and potential penalties for compliance failure and responding to regulatory demands, firms are becoming more analytical of regulatory legislation, organisations and effectiveness.

The regulatory bodies of the southeast lead the country in terms of efficiency, organisation and resourcing, but there are widely differing experiences: CETESB in São Paulo state is the best resourced agency (see Philippi, 1992; Crespo, 1993); Rio de Janeiro state (FEEMA) is under-resourced and influenced by the state's poor financial condition; Minas Gerais state (FEAM/COPAM) specialises in iron and steel regulation; and Espírito Santo is a small state with few large-scale industrial operations. Regulations with respect to the steel companies have in most instances been instituted within agreements – *Termos de Compromisso*, whereby the regulator and the firm have identified long-term programmes of priorities and investments.[7] This approach has been common during the 1990s and has been relatively successful since it identifies a time frame, and it specifies responsibilities, investments and actions, the regularity and efficiency of equipment and the on-going costs of monitoring and maintenance. Within the Agreements, the firms have greater flexibility in terms of regulatory compliance (although the agreed plan must be followed to avoid penalties), and the regulator can negotiate reasonable objectives.

Reaction to the agreements have not been uniform, however. Leading companies have taken positive steps towards environmental management and the meeting of emissions targets, but other companies have struggled to invest in these systems, and regulators remain critical. For example, FEEMA questions the commitments of CSN and its environmental performance, while Cosipa in São Paulo state (Cubatão) is closely monitored by the Santos branch of CETESB (see Fornari, 1991; Gutberlet, 1996b). In these two cases there is apparent disagreement between regulators and firms in terms of their environmental achievements and current performances. The difference in terms of agency is that FEEMA continues to struggle against low resources – which impedes inspections – while Cosipa is inspected very regularly by CETESB.

Apart from the construction of agreements and the nature of inspection and monitoring, an important difference is found in the levying of penalties against non-compliant firms. FEEMA penalisation of CSN has led to the firm having to invest these sums in *Termo de Compromisso* objectives (planned and agreed environmental improvements), whereas Cosipa receives none of its fines for tied investment (the penalties are directed to the state treasury). On the other hand, Rio de Janeiro state does not operate a finance initiative for environmental technology investment which São Paulo state does (PROCOP, administered by CETESB) (see Mello, 1990). This initiative is an attempt to move away from regulatory enforcement towards co-operation, incentives and motivation. These two examples reflect the differences in attitude towards environmental control and resourcing in the two states. In the case of Minas Gerais (FEAM/COPAM), a regulatory team focusing exclusively on ferrous metallurgy appears to have a strong relationship with industry and firms have responded well, for example, Usiminas and Açominas. Rather than steel-firm contamination, the agency has been more concerned with the use of charcoal and emissions from the pig iron, *guseiro* sector.

An effective regulatory system is dependent on legal support and a penalty system that creates a deterrent for companies while at the same time encouraging them to institute innovative practices and modern technologies. Currently, the Brazilian regulatory system is variable in terms of delivering deterrents and incentives, but it would appear that firms are responsive to regulatory pressures, although the starting points, the pace of change and the relationships between regulators and regulated are very different. Generally speaking, the relationship between regulators and industry has not been a good one. Much of this lies in the background of the heavy industries during the 1960s and 1970s under the authoritarian administrations when environmental controls went unheeded. The traditional criticism was that the regulators were anti-developmentalist, concerned only about pollution control rather than the social and economic issues associated with the changes, and inflexible. In return, the regulators believed that the industries were unlikely to adopt major environmental protection and control measures without strong state pressure and penalty systems.

The changes of the 1990s, with privatisation, opportunities for establishing *Termos de Compromisso* and greater environmental awareness within the industries (in terms of community, markets and legislation) have ushered in a new wave of regulation that has sought more flexibility where possible, and greater co-operation and assistance. This has been possible in most cases, but command-and-control has still been necessary where companies have been slow, or intransigent, in implementing necessary changes. More generally, the IBS (1990) argues that the steel industry has passed through four phases with respect to regulation: negation of the problem; reactive negotiation; tendency to co-operate and a proactive attitude; and the modern approach to accept the need to protect the environment. Different companies are at different stages, but the leading companies are increasingly aware of the need to reach stage four and be internationally competitive, hence the widespread application of ISO 14000 series measures.

Market considerations

Apart from the regulatory impacts on business, other 'drivers' in terms of improving environmental performance come from social and market pressures. The social pressures can be localised considerations, or social in terms of consumer behaviour and the impacts that changes in this behaviour can have on the firm. The latter is closely aligned with the pressures from the marketplace, and it is these which currently have greater impact than community pressures, although they may affect the attitudes of local regulators and sources of opposition. The variations across firms and their local opposition reveal that the community considerations are less clear than the pressures exerted by the regulatory authorities and the marketplace, by downstream client firms for instance. This variation can be seen in terms of the poor history of environmental performance by Cosipa and CSN despite their location in close proximity to residential areas, as opposed to CST which has had to respond to organised community action.

Market considerations relate to customer attitudes to the firm and the product and this is related to the profile of the firm. Environmental publicity is an important factor in this equation and is increasing in terms of importance. An extension of this market issue is the liberalisation of the economy and the domestic versus export market focus of the firm (the type of product and destination), also the impact of foreign relative to domestic investment. The cultivation of steel markets is a priority for producers, especially in the light of the low consumption per capita within the country and the expansion of capacity to meet new domestic demand (see Fig. 5.8), in particular the auto sector where flat-product producers have been developing a range of finished steels in collaboration with client companies.

For reasons linked to competitiveness based on inputs and product development, domestic producers have succeeded in keeping steel imports low at 2.7 per cent of apparent consumption in 1996 (2.1 per cent in 1992; 2.5 per cent in 1989 – IBS, various). Low iron ore and labour costs, allied with the sea transport costs for imports, make Brazil a market which is difficult to access. There is also no evidence of new foreign investment in the sector to produce domestically and circumvent the import obstacles. Much of this is a result of the relatively recent globalisation of the sector, although it is surprising that foreign companies did not take advantage of the privatisation process to purchase local firms and exploit the low domestic consumption rates. Explanations for this may lie in the collaboration of investment groups within the country, as well as the fact that most new international investment is taking place in the EAF sector rather than in large integrated plants.

Trade liberalisation prompted by the Plano Real has forced Brazilian firms to become more competitive, focusing on their productivity levels, labour reductions and better training. The new capital derived from privatisation has also allowed the application of new technologies and the decommissioning of older (more contaminating, less efficient) equipment. For the domestic market, there would appear to be little threat from non-Brazilian producers and few signs of impending foreign investment in the sector. Liberalisation has had more influence on those firms focusing on export markets. The

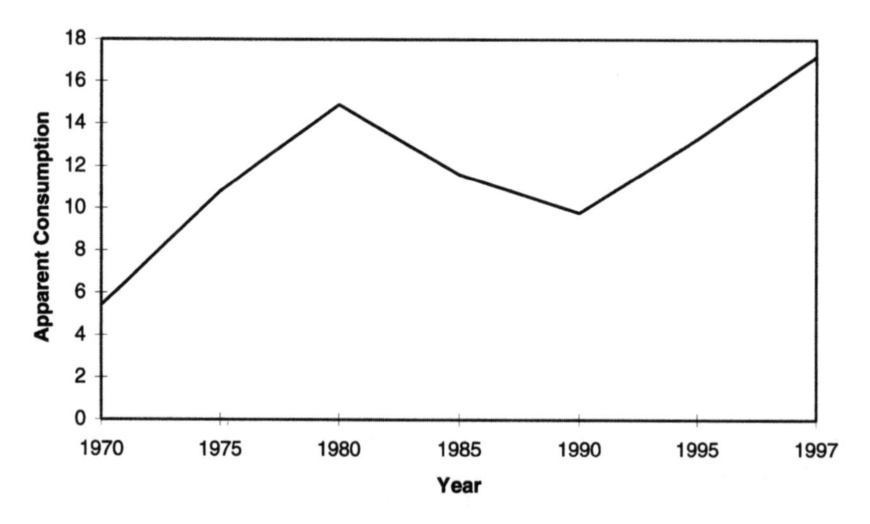

Figure 5.8 Apparent consumption of steel, 1970–97. Source: IBS (various years) *Anuário Estatístico*.

approach of these firms has been to produce low-cost steel, much of it semi-finished. The low production costs do not make the transportation costs prohibitive and allow these products to reach a range of markets. Semi-finished exports increased from 40.2 per cent of all exports by volume in 1986 to 65 per cent in 1998 (IBS, various; OECD, 1999); this contrasts with an average of 15–30 per cent ingots and semi-finished products amongst steel exports across the world during the 1990s (IISI, 1998). In terms of the environment, the export of products that have passed through the most pollution-intensive phases, but have not been processed to add value, suggests that these firms have a high environmental impact per unit value of product. This supports the debate that newly industrialising countries are havens for polluting processes, providing intermediate products for finishing in developed economies. The case of Brazilian leather and tanning is another example of this fragmentation of process stages.

As a factor in market access and sales, the environment does not rank alongside price and quality (ISO 9000); however, there is little doubt that considerations relating to the environment are increasing in their significance within corporate strategy. In Brazil, the surge in EMS certification has come mainly from firms that are internationalised in terms of capital and/or trade. The concern, real or perceived, is that developed economies may use environmental arguments to restrict trade in certain products or with certain firms. As a stamp of approval for their processes and products, Brazilian firms have sought international accreditation and association with good practice organisations; CST, CSN and Usiminas are members of the Brazilian Business Council for Sustainable Development. CST is Brazil's leading steel exporter (amongst the top three national export firms, 1994–7) and although the other two firms produce increasingly for domestic consumption, the value of their exports remains high: CSN has ranked amongst the top ten export companies since 1991; Usiminas during 1989–93 and 1995. A reflection of the rise in semi-finished exports is the emergence of Açominas into the top ten in 1997 (CST, 1997).

Environmental profile is considered increasingly important within marketing and sales strategies and it is likely that all the leading Brazilian firms will be ISO 14001 certified by 2001. Although certification is no guarantee of environmental performance, since it concentrates on organisation and procedures, the existence of an EMS would suggest that a mechanism for continuous improvements is in place; its main impact is in terms of recognition of attention to environmental management. ISO 14001 is considered to be a useful marketing tool for export products as international clients become more sensitised to environmental issues, such as life cycle analysis of products and the ability to extend firm-level environmental policies along the supply chain.

Some of the same issues apply to domestic market-oriented firms. In Brazil, the demand from the auto sector is considered to be the most important for the steel sector. Since the auto sector on an international scale is having to respond to pressures for 'cradle-to-grave' management of cars, lighter body cars and 'green'-fuelled cars, the steel sector is having to meet tighter controls passed down the supply chain. Environmental management and performance is an important aspect of these controls. Although steel is most often one step removed from the retail sector, the demands from consumer organisations and environmental organisations are becoming more important. Since the leading auto companies in Brazil are international firms, Brazilian steel companies are responding to these international pressures.

ISO 14001 is considered to be a qualification that is widely accepted by client firms and that demonstrates a commitment to responding to environmental issues. Although it in no way indicates a change in environmental performance, ISO 14001 certification is the most readily available tool for firms to adopt 'green credentials'. Overwhelmingly, the drivers are *perceived* market pressures, that firms *will* increasingly ask for certification, as with ISO 9000, and that it *may* help products entering particular countries or trade blocs as environmental issues have increasing impacts on trade conditions and preferences. The added benefit that the certification can be used to satisfy worker and community groups' concerns about environmental issues is also important, despite the lack of environmental performance criteria. Currently, evidence of these drivers is weak, but Brazilian steel firms are planning ahead and regard ISO 14001 as an 'eco-label' that can accompany the ISO 9000 'quality label'. The accreditation and certification agencies are also influential in this process, advocating its benefits by promoting these uncertainties in international market and client firm demands.

Variations in firm approach

Amongst the five firms under consideration, it is possible to make a number of observations relating to environmental performance and the drivers of environmental performance improvements. The first is that one can identify three different approaches. The first is the case of Usiminas which has taken a proactive approach to environmental protection and early ISO 14001 certification. It has set challenging emissions' reductions targets and also prides itself on its community environment programmes: the Projeto da Vida and Projeto Xerimbabo. Despite its large scale, the firm works well with the regulator within the joint Agreement and has little opposition from the residential area near to the plant; 'greening' and resident workers may explain the latter.

The second approach is less proactive but is active and attempts to stay ahead of regulatory developments. Both CST and Açominas are in this category and both are semi-finished steel producers and of 1970s construction. The fact that they are export-oriented firms is a factor in their approach to environmental management since there is greater awareness of international issues around environmental protection for their international clients, particularly those in developed economies. Nevertheless, evidence of direct impacts remains limited, and neither of these semi-finished export producers notes that their clients are currently requesting information regarding environmental performance. For these firms, their ability to remain compliant with national regulations is facilitated by the fact that many environmental protection systems were constructed from the outset on these sites. A dimension that separates the firms is their location. Açominas has an isolated site with no close residential areas, chosen for raw materials and water availability reasons. CST on the other hand was located for port facility and sea-water availability factors and is close to the state capital. Consequently, it has been pressurised by the city authorities (as has its neighbour, CVRD) and has had to work hard on a local level with various organisations (local government, university, residential groups and media) to reduce environmental criticism.

The aforementioned approaches reflect an increased environmental awareness over time. The third approach reveals a history of poor environmental awareness, and developments that have been driven by regulations rather than self-interest. CSN and Cosipa both fit this category but with different circumstances. The case of CSN is important in that it has been a flagship national firm since the 1940s. Its role in successive national development strategies and its significance within the economy of Rio de Janeiro state have provided it with considerable flexibility. The regulator has fought a long battle with the firm in order to improve its environmental record, but it has had limited success. The plant's proximity to residential areas and its influence within the river Paraíba do Sul catchment have led to improved regulatory enforcement, particularly since privatisation, but a legacy of decades of contamination and the age of some existing equipment will ensure continuing high environmental costs in the short term. Of particular significance in the case of CSN is the weakness of the regulator FEEMA, which is under-resourced and cannot carry out its monitoring and inspection activities as it would wish. The outcome is antagonism and difficulties in carrying through the developments agreed upon in the *Termo de Compromisso*.

For Cosipa, located in Cubatão (São Paulo state), circumstances are different although there are similarities in terms of its history of poor environmental performance and the very recent shift towards compliance and increased investments in environmental technologies. The principal differences are in terms of location and regulatory control. Cubatão has a high profile in national environmental consciousness due to its concentration of 'pollution intensive' industries and the geography of the area which intensifies the emissions' impacts. The city's environmental issues have become highly politicised and in spite of various programmes to reduce emissions, legal claims have been made against the firms on health grounds. Since the early 1980s, CETESB has been very active in the area and monitors developments very closely through its Santos branch. Within the Agreement with Cosipa, new environmental investments have been forthcoming and fines are still used when targets are not achieved. The firm–regulator relationship is very close and CETESB, which is relatively well resourced, is able to assert itself and force Cosipa towards statutory limits.

In both the CSN and Cosipa cases, the drive towards compliance is still not complete and is being confronted with predominantly technological developments. There is no strong environmental management emphasis, despite a highly formalised approach to move through the stages of auditing and procedural changes to achieve ISO 14001 certification, and the firms appear to be considerably less advanced than others in terms of taking on board the implications of environmental awareness for neighbours, clients and markets.

In terms of potential risks relating to location, clear differentiations can be made between the isolated site of Açominas, the close residential areas of CSN, and the industrial concentrations and residential impacts of Cosipa within Cubatão. Reporting of information, community support and stringent regulatory controls are imperative within a setting of residential proximity, taking precedence over the longer-term emissions, discharge and by-product impacts to waterways and vegetation. Firm location is highly influential in terms of what limits should be achieved in order to ensure minimum health and natural environment impacts. Across Brazil's states, regulatory agencies are sensitive to local dynamics, but the suggestion is that there is a long way to go for some firms (even if compliant under existing legal prescriptions).

Although the approach to environmental issues shapes management policies, the financial backing to carry policies through is vital. The difficulty with assessing environmental investments is that the available figures are indicative rather than comprehensive. In their accounts, firms most often restrict this category to capital investments in equipment, rather than incorporating the on-going operating, monitoring and management costs of these technologies, or the costs of extension such as community programmes. An important caveat in terms of assessing environmental investment patterns is that there has been an incremental rise from the 1970s to late 1990s; therefore, the longer-term assessments offer the best insights. IBS estimates that 5.9 per cent of total investment during the 1994–2000 period will have been committed to environmental issues; this rises to almost 12 per cent if energy investments are included (see Table 5.7). For instance, a firm may have invested as much in one year during the 1990s as it did in five years during the early 1980s. One can detect clear differences between the firms based on their own data; if one adds the dimension of investment per unit of output, the lack of investment in CSN becomes particularly notable. Despite being built with basic environmental equipment, the investments of CST and Açominas remain higher than those of CSN, and even the CSN's most recent investment plan appears to undervalue the need for environmental investments (6 per cent against Cosipa's 23 per cent). What is most apparent is that Usiminas has the best environmental performance to date and this has been achieved by targeting investments in that area during the 1990s. Connected

Table 5.7 Total investments by area, 1994–2000

Area	*US$ million*	*%*
Energy generation	626	6.0
Reduction and steelmaking units	3,360	32.3
Rolling mills	4,159	40.0
Environment	613	5.9
Modernisation, automation and research	1,643	15.8
Total	10.401	100

Source: IBS (2000).

to these investments is a management structure for the efficient operating and control of these new investments in order to ensure continued environmental performance improvements.

The 'greening' of Brazilian steel

From the early 1990s, Brazil's steel sector has been privatised, modernised and expanded. This has been part of a national strategy guided by the Plano Real, and a response to rising long-term domestic demand during the 1990s following a steady decline in the 1980s. Domestic steel consumption rose by an average of 8.2 per cent per annum during the period 1990–7 as production and imports rose and exports remained relatively stable, and per capita steel products' consumption rose from 60.9 to 95.8 kg/cap (IBS, 1999). Privatisation has stimulated price competitiveness, rationalisation of production, product development, quality, raw material and energy supply security, and also environmental investment and management. Although the international market is competitive, advantages in labour costs (although not in labour productivity necessarily) and raw material availability (particularly iron ore, and charcoal for some firms) provide the sector with a strong competitive basis. A recent consultancy report on competitiveness for IBS reveals the areas in which further progress is required. Environmental investments rate highly relative to what one might have expected 10 years ago, but the area of most pressing need is for investment to increase productivity, followed by personnel training (see Table 5.8).

Firms should be given credit for rapid environmental developments since the 1990s, but the starting point was a low one and considerable work remains to be done. Much of the improvement has been in the application of readily available technologies such as continuous casting and the recycling of production gases for energy savings. These cleaner technology developments should be welcomed, but they have been slow in coming for most firms compared with developed economy firms.

Table 5.8 Competitiveness impacts by activity

Activities	*Impacts*[a] *(1 = low; 4 = high)*
Investments	
expansion	1
productivity	4
quality	2
environment	2
Product development	2
Personnel training	3
Market development	
domestic	1
international	2

Source: Booz-Allen and Hamilton (1996).
Note
a Positive impacts on competitiveness.

The *Termos de Compromisso* approach to regulating firms is favoured over a more command-and-control regulatory style; however, the flexibility of this approach and the different uses of penalties from non-compliance lead to considerable variations in terms of firm experiences of regulation in different locations. The resourcing and efficacy of inspections and monitoring is allied to this point, and is influential in terms of firm behaviour. Beyond regulations, it is the firm that will determine its approach to the environment, either positively via executive prioritisation (in the case of Usiminas) or reactively (CSN). These decisions will be made in response to market access, the need to work with stakeholders to improve local environments (for public relations and moral obligation reasons) and broader firm environmental consciousness. The enthusiasm for ISO 14001 amongst steel companies is evident and it is apparent that progress is taking place in systematisation, management structures and procedures, and awareness via training. A major caveat is that ISO 14001 does not guarantee environmental performance improvements. For these to take place, regulators and firms must audit processes and reach agreed and practicable targets. These are evident in the *Termos de Compromisso*; however, the penalties for not meeting agreed targets must also be evident and effective.

The liberalisation of the Brazilian economy has had less impact on the steel sector than on others. Levels of foreign investment remain relatively low and in terms of trade, the price advantages derived from low labour costs (in spite of low productivity) and iron ore availability outweigh the 'Brazil Cost', the costs of buying coal on the international market and the transportation of exports to markets in the USA (24 per cent, 1996), Latin America and east and southeast Asia. The potential steel demand in the country also keeps the sector buoyant and has led to planned expansion, against a backdrop of global overcapacity. Environmental protection has been influenced by international developments. The ISO 14000 series and the Business Council for Sustainable Development initiatives have been taken up by firms in order to keep their options open. There is concern that international markets may be increasingly affected by environmental protectionism, and that these mechanisms act as safeguards although firms have not yet experienced downstream pressures from clients. Despite the increasing focus on domestic market demand, the large firms are aware that they need to remain flexible. On the other hand, the Gerdau group (which operates four plants in the country) produces almost entirely for the domestic market and has no intentions of developing an EMS for certification, choosing instead to focus on eco-efficiency gains; this may be typical of other Brazilian firms focusing exclusively on the domestic market.

A final consideration is the recognition of two distinct phases of sectoral development. Pre-privatisation one can say that companies polluted freely in the pursuit of high output figures. Post-privatisation, the environment has been emphasised as a strategy alongside the priorities of modernisation and increased competitiveness. There are considerable variations between companies and there are still opportunities for energy consumption and emissions' reductions, and greater recycling to achieve a more manageable, sustainable and 'accepted' industry. State regulations, communities and markets will drive this process, but it is only via co-operation and realistic objectives that the sector will be able to incorporate environmental considerations into strategic production and commercial considerations. The goal of *aço verde* ('green steel') is a long way off, but when one considers how environment protection has been incorporated via investment and better management since 1980, there should be optimism for innovative developments and continued environmental performance improvements in order to produce steel efficiently and without lasting damage to human health and natural environments.

Notes

1 This material was gathered between October and December 1997 as part of an EC-funded project titled 'Environmental Regulations, Internationalisation of Production and Technological Change' (ENVA-CT96-0235). I would like to thank the IBS, the steel firm managers, the regulators and the sectoral experts who contributed to the study.

2 As a significant contributor to atmospheric contamination in the Greater Vitória region (with its neighbour, CVRD), CST has sought to reduce its emissions and works in co-operation with a local municipality agency and the state university (CST, 1997).

3 In the case of Usiminas, investments in environmental technologies totalled US$81 million to 1976, but this figure increased to US$170 million by 1989. The balance between pollution media was as follows: 61.8 per cent atmospheric; 37.4 per cent water; 0.6 per cent noise; and 0.2 per cent soil. During the 1990s, the

accumulated environmental investment rose to US$356 million (1996) (Campos Soares, 1990; Usiminas, 1996b).

4 Between 1995 and 1998, 32 pollution control systems were expected to have been put in place, concentrating principally on air emissions. The plans for 1997/98 included centralised particulates' monitoring, the first step towards ISO 14000, which is pollution reduction and control, and improved training with the assistance of CETESB (environmental training was initiated in 1996).

5 Jose Maciel (1993) cites the case of Açominas and the waste that is generated from the gas-cleaning system in the steelmaking process. It is calculated that the waste is 59 per cent iron and 2 per cent coal and Açominas generates on average 7000 tonnes/month (t/m), which is a loss of 4141 t/m of iron and 158 t/m of coal dumped in industrial waste deposits. Since the cost to deposit this material is approximately $5 per tonne of waste, actual losses are US$35,000 per month. The iron can be recycled in the process and there is also research into commercialisation of the whole range of by-products.

6 In the case of Usiminas, there are a range of activities designed to promote environmental consciousness in the company and within the community (Campos Soares, 1990): *Internal* – internal technical seminars; environment week; 'tree day'; operator demonstrations; green areas programme. *External* – plant visits; assistance to community initiatives (clubs and associations); first-grade environmental education assistance.

7 CSN signed an agreement with FEEMA (the Rio de Janeiro regulatory agency) in 1994 to resolve its environmental problems before May 1994. The agreement included 325 items. The company claims that 80 per cent of the items in the agreement have already been achieved. For CST, the *Termo de Compromisso* was drawn up in September 1990, outlining 40 items which had to be accomplished by 1994. In October 1994, 45 more items were added in order to continue the environmental performance improvements process (Guimarães, 1996). CST invested US$60 million to achieve the 1994 demands. For Usiminas, the *Termo de Compromisso* with COPAM (Minas Gerais regulatory agency) of 1990 had the objective of instituting adequate operational facilities to meet environmental regulations. This was renegotiated in 1994 and 1996 with 37 actions to be undertaken with an associated cost of over US$170 million (Usiminas, 1996b).

Bibliography

Almeida e Silva, R.de (1995) 'A Gestão Ambiental na CST', *in Encontro Regional, Meio Ambiente na Industria Siderúrgia 5* (IBS/ILAFA), 2-16 STI-1/1-15.

Campos Soares, R. (1990) 'Gerenciamento Ambiental: A Visão e Experiencia da Usiminas', *Anais I: Simposio Nacional de Gerenciamento Ambiental na Indústria*. São Paulo: ABIQUIM, 1–21.

Centro Brasileiro de Qualidade, Segurança e Productividade (QSP) (1994) *ISO14000: Sistema de Gerenciamento e Certificação Ambiental*. Vol. I: *Anais do Primeiro Seminario*. São Paulo: QSP.

Cetesb (1995) *Seminário ISO14000 Certificação Ambiental*. São Paulo.

Claudio, J. R. (1994) 'A importância das certificações ambientais para as empresas', *Saneamento Ambiental*, 5(28): 20–3.

Companhia Siderúrgica de Tubarão (CST) (1997) *General Information*. Serra, ES: CST.

Companhia Siderúrgica de Tubarão (CST) (n.d.) *CST's Environmental Profile*. Serra, ES: CST.

Companhia Siderúrgica Nacional (CSN) (1996) *Annual Report*.

Companhia Siderúrgica Nacional (CSN) (1997) *Quarterly Report – ITR*, 30 September 1997 and 1996.

Companhia Siderúrgica Nacional (CSN) (1997) 'A CSN e o Meio Ambiente', *Metalurgia & Materiais*, July: 38–9.

Companhia Siderúrgica Nacional (CSN) (1997) *General Information*.

Cosipa (1997) *A Gestão do Meio Ambiente na Cosipa: Retrospectiva e Perspectivas (Sumario Executivo)* (Cubatão, Cosipa).

Costa Ferreira, L. da (1996) 'A Política Ambiental no Brasil', in G. Martine (organiser) *Populacão, Meio Ambiente e Desenvolvimento: Verdades e Contradições*. Campinas: Editora da UNICAMP, 171–82.

Crespo, R. (1993) 'Aos 25 Anos, a Cetesb traça novos rumos', *Saneamento Ambiental*, 4:24, 16–23.

Dantes, M. A. and J. Souza Santos (1994) 'Siderúrgia e Tecnologica (1918–1964)' and 'Siderúrgia e Tecnológica (1964–1980)' in S. Motoyama (org.) *Tecnologia e Industrialização no Brasil: Uma Perspectiva Histórica*. São Paulo: Editora UNESP, 209–232, 233–50.

EC (1996) *Co-ordinated Steel-Environment Programme*. Luxembourg: EC.

Fornari, M. C. (1991) 'Cetesb melhora controle da poluição', *Saneamento Ambiental*, 2:13, 20–26.

Freitas Alvim, D. R. and T. Morimoto (1995) 'Auditoria Ambiental Interna Experiência da CST', in *Encontro Regional, Meio Ambiente na Indústria Siderúrgia 5* (IBS/ILAFA), STII-4/1-12.

Gama Torres, H. da (1996) 'Indústrias sujas e intensivas em recursos naturais: importância crescente no cenário industrial brasileiro', in G. Martine (organiser) *Populacão, Meio Ambiente e Desenvolvimento: Verdades e Contradições*. Campinas: Editora da UNICAMP, 43–68.

Gazeta Mercantil (1997) *Balanço Anual 1997*.

Gerdau Group (1996) *Relatório Anual*.

Guimarães, M. B. (1996) 'O Controle Ambiental: Relaçionamento Governo,Comunidade, Indústria e Universidade', *Encontro Regional, Meio Ambiente na Indústria Siderúrgia 5* (IBS/ILAFA), STI-5/1.

Gutberlet, J. (1996a) *Produção Industrial e Política Ambiental: Experiênçias de São Paulo e Minas Gerais*. São Paulo: Fundacão Konrad-Adenauer-Stiftung.

Gutberlet, J. (1996b) *Cubatão: Desenvolvimento, Exclusão Social, Degradacão Ambiental*. São Paulo: EDUSP.

IBS (various years) *Anuario Estatístico da Indústria Siderúrgica Brasileira*.

IBS: Comissão de Assuntos Ambientais (1990) 'Gerençiamento Ambiental na Indústria', *Revista Engenharia Ambiental*, 3:9, 23–9.

IBS (1993) *Gestão Ambiental no Setor Siderúrgico Brasileiro*. Rio de Janeiro: IBS.

IBS (1995) *Environmental Management in the Brazilian Steel Industry*. Rio de Janeiro: IBS.

IBS/Booz-Allen & Hamilton (1996) *A Siderúrgia Brasileira: Competitividade*. Rio de Janeiro: IBS.

IISI (1998) *Steel Statistical Yearbook*. Brussels: IISI.

Maciel, J. C. (1993) 'O Custo da Poluição' *Anais, Simposio sobre Controle Ambiental na Siderúrgia 6.* Rio de Janeiro, June.

Mannesmann Florestal Ltda (1996) *Technological Development: Support for Sustainable Forests.* Curvelo: MG.

Maumédio de Paulo, M. (1992) 'O Controle de Resíduos Industriais e Subprodutos na CST', *Metalurgia & Materiais*, 48:408, 476–83.

Medeiros, J. X. (1995) 'Aspectos Econômico-Ecológicos da Produção e Utilização do Carvão Vegetal na Siderúrgia Brasileira', in P. H. May (org.) *Economia Ecológica: Aplicações no Brasil.* Rio de Janeiro: Editora Campus, 83–114.

Mello, S. R. de (1990) 'Procop financia controle ambiental', *Saneamento Ambiental*, 1:5, 12–5.

Ministry of Housing, Spatial Planning and the Environment, The Netherlands (1997) *Dutch Notes on BAT for the Production of Primary Iron and Steel.* The Hague.

Morandi, A. (1997) *Na Mão da Historia: A CST na Siderúrgia Mundial.* Vitoria: EDUFES.

Neder, R. T. (1996) 'O Problema da Regulação Pública Ambiental no Brasil: três casos', in L. Da Costa Ferreira and E. Viola (orgs) *Incertezas de Sustentabilidade na Globalização.* Campinas: Editora UNICAMP, 217–40.

OECD (1999) 'Steel Market Development in 1998 and Prospects for 1999', *OECD News Release*, 10 May.

Oliveira, M. A. de and F. M. A. Tourinho (1996) 'Estudo da Productividade no Sector Siderúrgico Mundial: Uma Comparação Possivel', *51 Congreso Anual da ABM.* 5–9 August, Porto Alegre, 835–1.

Philippi, A. (1992) 'Recursos para o Controle da Poluição Industrial no Estado de São Paulo', *Anais do Seminario Internacional, Industrialização e Meio Ambiente: Reciclagem do Lixo e Controle da Poluição.* Vol. II. São Paulo: INTER/CETESB/FUNDAP, 211–21.

Pinheiro da Silva, S. (1992) 'Siderúrgia Brasileira: situação atual, tendências e perspectivas', *Metalurgia & Materiais*, 48:408, 488–95.

Porter, M. and C. van der Linde (1995) 'Towards a New Conception of the Environment–Competitiveness Relationship', *Journal of Economic Perspectives*, 9:4, 97–118.

Usiminas (n.d.) *Certificação Ambiental: Experiênçia e Desafio da USIMINAS/DNV*

Usiminas (n.d.) *USIMINAS: Um Projeto de Vida.*

Usiminas (n.d.) *Meio Ambiente.*

Usiminas (1996a) *Gerenciamento Ambiental.* Belo Horizonte: Usiminas.

Usiminas (1996b) *Cartilha do Meio Ambiente.*

Usiminas (1997) *O Gerenciamento Ambiental na Usiminas.*

6

ECONOMIC LIBERALISATION AND THE ENVIRONMENT – A CASE STUDY OF THE LEATHER INDUSTRY IN BRAZIL[1]

Jan Thomas H. Odegard[2]

Introduction

In the 1990s two major political global processes have had their 'take-off': efforts to liberalise the world economy, and efforts to protect the local and global environment. An interesting point is that the two processes represent forces that tend to pull in different directions. While the aims of the Uruguay Round of the GATT negotiations and establishment of the World Trade Organisation (WTO) in 1994 were national deregulation and the freeing of market forces, the aim of Agenda 21 following the Earth Summit in Rio de Janeiro in 1992 was to introduce regulatory measures to prevent further pollution and exploitation of natural resources by human activity. Generally stated one could say that many of the activities that the WTO is aiming at liberalising are the same ones that environmental regulations are trying to get under control.

In order to understand the interaction between the two phenomena better, it is fruitful to study cases of industries and economic activities that have been subject to both processes. The purpose of this chapter is to illustrate this interaction by looking at the changes that have taken place in the Brazilian tanning industry in the 1990s. Until the mid-1980s, Brazil was one of the most protectionist countries in the world, often referred to as having an inward-oriented development model. Self-reliance and independence were important strategic goals. The start of the democratisation process in 1985 also triggered efforts to liberalise the economy and trade. These efforts did not have a significant impact until the neo-liberal president Fernando Collor de Mello took office in 1990. Collor de Mello was also the host to the

Earth Summit in Rio in 1992. He made efforts to improve the poor environmental image of Brazil and the country's environmental performance. The result has been two parallel processes, one of increased economic liberalisation and another of stricter environmental regulation.[3]

The tanning industry has three distinct features that make it a good case for analysing the two processes. First, tanning is an industry with a high pollution load. Second, leather is a world market commodity, but supply and demand vary independently of one another. The supply of hides depends on the demand for meat, while the demand for leather varies with fashion trends and climatic variations. Finally, leather can be divided into three products, each the result of a different processing stage. The production stages of 'wet–blue', 'crust' and 'finished leather'[4] each involves a distinct level of value added, capital intensity and pollution. These distinct characteristics result in different influences on each product stage by national and local incentives and regulation, resulting in different patterns of location and interaction with other industries.

This chapter consists of five parts. The first is a presentation of the structural changes that the Brazilian tanning industry has passed through during the last decade. Then follows a presentation of the leather production process and its pollution load. The third part concerns the change in environmental regulation in Brazil, both in terms of the law and its enforcement. The fourth section is an attempt to answer the specific questions of the extent and the ways in which economic liberalisation has influenced the environmental performance of the tanning industry. This is followed by a final consideration of the apparent trend for Brazilian tanners to specialise in the export of a low value added and pollution intensive product.

Restructuring of the tanning industry in the 1990s

At the end of the 1980s Brazil had become virtually self-sufficient in leather and leather shoes, and was one of the main exporters of these items in the world. During the last two decades Brazil produced and exported about 5 per cent of all world leather. From the late 1960s until 1986 the output of light bovine leather[5] almost doubled, and from 1986 to 1994 output grew by another 40 per cent (FAO, 1997). By the early 1990s, Brazil was the largest producer of light bovine leather in Latin America and the fifth largest producer in the world after South Korea, Italy, China and India. Production was carried out exclusively by domestic tanneries. The number of tanneries reached a peak in the mid-1980s with more than 700 companies employing over 70,000 people. Together with about 4000 shoe manufacturers (highest estimates), the leather industry employed, directly and indirectly, about one million people. More than 95 per cent of the bovine raw hides used were Brazilian, due to the fact that Brazil has one of the largest cattle stocks in the world.

During the 1990s there have been six main structural changes in the tanning industry. First, the industry has passed through a process that might be called *regressive restructuring*.[6] The number of companies has declined sharply since the late 1980s. Between 55 and 65 per cent of the companies have gone out of business and many of the surviving companies are heavily in debt (Gazeta Mercantil, 1998). There were about 250 working tanneries in early 1998, of which only 20 were doing well according to representatives from the industry. The general impression is that the crisis has not yet passed. The industry today still has excess capacity of approximately 30 per cent, and there are frequent bankruptcies. Brazilian tanning experts are seriously underemployed, and more than 200 have emigrated to China alone.

The second change is a greater concentration of production capacity into large groups. The stronger tanneries buy the bankrupt or weak companies and create scale advantages through accumulated production capacity. Already in 1993, 11 per cent of the tanneries accounted for 70 per cent of total turnover, while 65 per cent of the tanneries were of the artisan type with less than 10 per cent of production. The large groups also include foreign investors that go in as owners or rent surplus production capacity from domestic tanneries. Italian tanners in particular are on the offensive to gain production capacity, and estimates among the tanners in Brazil are that the Italians own companies with 10 per cent of the total production capacity for wet–blue, and have an ownership interest or rent production capacity in companies that amount to another 30–40 per cent of the total production in the Brazilian tanning industry.

The third change has been the growth of the so-called 'briefcase tanners' (originally *Curtumes de Pasta*). These are usually individuals that use their tanning experience, networks and favourable access to capital to utilise the excess production capacity of the decapitalised and indebted tanners. The purpose is to maximise profits through the organisation of a production chain. The briefcase tanners are a phenomenon that can also be observed in other countries with similar conditions, such as the Czech Republic. They are said to emerge in periods of restructuring and decapitalisation when they can make quick profits, while they take on other roles (such as running tanneries or dealing in hides) in periods of stability.

The fourth structural change is that the tanning industry has become increasingly export oriented. While 20 per cent of production went to direct exports in 1987, this grew to 27 per cent in 1990 and 50 per cent in 1997. Indirect exports, mainly as shoes, decreased from 39 to 26 per cent and then to 22 per cent over the same period.

This growth in exports is closely connected to the fifth change, a 32 per cent growth in the availability of domestic hides from 1990 to 1997. This is a result of almost doubling the slaughtering of cattle from 16 million heads in 1984 to 29 million heads annually in 1997. This development was a result of the increased intensity of cattle raising for meat. It is important to point out that this was not a development driven by demand for hides, as the hide only makes up 10–15 per cent of the slaughter value. In 1997 Brazil had 160 million cattle and produced 29 million hides annually, while the USA produced 40 million hides from 100 million heads of cattle. This shows that there is still a substantial potential for increasing the rate of slaughtering, and the trend of growth in the annual volumes of hides is likely to continue.

While the export rate of other leather products has remained stable or declined since 1993, all the growth of available hides have gone into the production of wet–blue for direct export, and this accounts for 90 per cent of the total growth in leather exports. In 1997, 72 per cent by weight and more than half the earnings from leather exports were due to wet–blue. This development is shown in Table 6.1.

A fundamental characteristic of wet–blue production is that whereas 15 per cent of the value added in the leather production chain is generated during the wet–blue process,[7] more than 80 per cent of the pollution occurs in this stage of production (Miljøministeriet, 1992). Moreover, the wet–blue process is financially less risky, because the wet–blue can be used for almost any end-product and is more easily traded to downstream customers. It allows for economies of scale and is neither particularly knowledge nor labour intensive. It also has a shorter production cycle, so that less capital is bound up in high-value raw material in this part of the production process.[8]

A sixth structural change is the relocation in two separate waves of tanneries to the interior central eastern and northern regions of Brazil, a process called 'interiorisation'. The first relocation of tanners started due to the movement of cattle and slaughterhouses approximately 20 years ago towards the central regions of the country. This upstream restructuring was a consequence of the price increase on pastureland in the industrialising southern regions, combined with significant incentives for cattle farmers to relocate to the less developed west central and northern parts of the country. The consequence has been that while more than half of the cattle were located in the

Table 6.1 Leather exports from Brazil by stage of production (in millions of hides)

Year	*1990*	*1991*	*1992*	*1993*	*1994*	*1995*	*1996*	*1997*
Salted hides	0	0	0	0	0	1	1	1
% share of export	0	0	0	1	1	6	6	4
Wet–blue	4	4	5	4	4	8	10	11
% share of export	59	61	62	51	57	69	66	72
Crust	2	1	1	2	2	1	2	2
% share of export	23	21	17	25	21	12	11	12
Finished	1	1	2	2	2	2	2	2
% share of export	18	18	21	23	21	13	13	13
Total export of leather	7	6	8	8	8	12	15	16

Sources: Abicouro (1997); ABQTIC (Associação Brasileira dos Químicos e Técnicos da Indústria do Couro) various years.

three southern states in the 1970s, only 17 per cent remained in 1997. In 1994 the highest concentration of cattle is located in the west central states (34 per cent). Increasing incentives directly to the industry to move to the new cattle states further motivated the relocation of tanners. Many of these incentives continue to exist, and include large tax reductions, cheap credit, land and equipment. Wages are lower, and environmental costs are also lower owing to cheaper land to build space-demanding treatment facilities and less strict regulations compared to those of the southernmost states. The last point is no less important as the majority of the new tanneries in the west central areas produce wet–blue (so-called *bluseiros*).

A second and more recent wave of relocation occurred in connection with the 'interiorisation' of the shoe industry. In the mid-1980s Brazil was the fourth largest producer of shoes in the world, following China, Italy and India. More than 500 million pairs were produced annually, of which 25–30 per cent were exported. The majority of shoe production was located in the industrial districts of Novo Hamburgo in Rio Grande do Sul and Franca in the state of São Paulo. The interiorisation of the shoe industry, the main customer of the tanneries,[9] has taken place mainly during the last five years. This development is related to the general restructuring of the shoe industry over the last decade, a consequence of the need to reduce costs to meet increasing and cheap imports from Asia both in the formerly protected domestic market, and export markets in the USA and Europe. Chinese shoes were offered to the market at average prices 35 per cent below Brazilian. As a consequence, Brazilian shoe production was almost halved from the peak year of 1986 to 1994, when the level was down to that of two decades earlier (FAO, 1996). Exports dropped from 155 million pairs in 1989 to 120 million pairs in 1992, which was below the 1984 level.[10] The situation became so threatening that the Brazilian government deemed it necessary to raise the import tax on the most vulnerable categories. Although the shoe industry has recovered somewhat in recent years, the turnover has stagnated (at US$1300 million) (World Leather, 1997/1998). In 1996, 580 million pairs were produced, which was 20 per cent below the production capacity of the industry.

By relocating to the northern regions, the companies take advantage of significant state incentives and substantially lower wages compared to the traditional shoe-producing districts in the south. In the northern areas people are poorer and less unionised. They often have access to farmland to supplement their income with home-grown food. A majority of the shoe companies are small, labour intensive and artisan, and do subcontracting for the handful of dominating conglomerates. Hence, the shoe companies are footloose, and relocate easily. Although this study found no detailed documentation on *tanners* relocating their production facilities northward in great numbers to be closer to the shoe companies, there has been a tendency for already established tanners in the northern and west central states to expand their production.

Tanning production process and its wastes

Tanning is the process of converting raw animal hides into leather. Leather is a unique and flexible material combining strength with softness and resistance to decay. It warms, protects, breaths, feels comfortable and has a pleasant appearance. Leather has been made by humans for thousands of years, traditionally using manual power and simple tools to clean the hides, while tanning was done by soaking the hides in a solution of water and the bark of certain trees for several months. It has always and still is being used in clothing, shoes, furniture, book covers and fancy-wear (purses, belts, etc.) and increasingly for seats in transport units. Large-scale industrial production of leather became possible in the twentieth century with the introduction of chemical tanning agents (mainly chrome) and machinery that allowed for higher scale and automated much of the tanning process. Even so, tanning still remains artisan today as the hides vary greatly in shape, thickness, composition and surface qualities, impeding complete automation as each hide can react differently to the chemicals and mechanical processes applied.

Today, the tanning process is made up of roughly a dozen chemical and mechanical processes. Initially, hair, dirt and flesh are removed. The collagen fibres which make up the hides are then treated chemically to make the leather strong and resistant to biological decay. The surface and body of the leather can then be treated in a multitude of different ways to obtain the colour, look, feel or other qualities required by the customer. For analytical purposes the tanning process is here divided into three phases: the wet process; retanning and finishing. The wet process includes the cleaning, basic tanning and horizontal splitting of the hides. The cleaning process and tanning take place in large drums (2–5 m in diameter) where hundreds of hides are soaked in chemicals and water for a day or two. This requires large quantities of water, which results in a wastewater cocktail of tanning agents, acids, fungicides and salts, and generally high BOD[11] and COD values. This requires various processes, as the hides have to be neutralised before tanning. If the tannery treats the wastewater, it has to deposit the sludge containing chrome. Finally, the still wet and now blueish hides (from the chrome, therefore called 'wet–blue') are *split* horizontally and then *shaven* to obtain even thickness and *trimmed* to get a more even shape. Shavings and trimmings are also solid wastes containing chrome, and should be deposited. Even if the wet–blue hides are wet, the treatment with chrome and fungicides makes them resist decay so that they can be transported and stored for several months.

The second phase is the retanning process. Here, the tanning process is more specialised and colour is added to the entire hide. Organic and chemical remnants are once again released to the wastewater, as well as dye, tanning agents and fats. The hides are then mechanically dried. Often the top surface is buffered (smoothing of the grain surface by mechanical sanding), which results in leather dust containing chrome that should be deposited.

The third phase is finishing, where paint and patterns are applied to the leather surface. The wastes from this process are mainly liquid and solid residues of finishing solutions.

The accumulated waste from the tanning process is more than 700 kg of solid waste for each ton of fresh salted hide (UNIDO, 1997). For each hide, large amounts of water are used that can contain hundreds of different chemicals. With conventional finishing, solvents are released to the air, but the introduction of water-based solvents has reduced this source of pollution. Even so, tanning continues to be one of the most pollution-intensive industries if left unchecked, which is the main reason why in many countries it has been one of the first industries to be targeted for environmental regulation.

Increased environmental regulation

Deregulation of trade and the stabilisation of the currency are important causes of the changes in the tanning industry. Even so, there is reason to believe that the environmental regulations introduced in Brazil in the 1990s have also had an impact. Until the mid-1980s the law on environment was fragmented and weak, and there is a consensus among the informants that there was almost a total lack of environmental awareness in general in Brazil. Environmental regulation and enforcement were to a great extent non-existent and had limited influence on the tanning industry as a whole. The new constitution of 1988 introduced important elements to improve the protection of the environment. Most of these proposals remained good intentions, as follow up was limited owing to political and financial turmoil. Only in the most extreme cases of contamination, such as areas with high concentration of tanneries, were efforts made to clean up. One such place was Estancia Velha (located in the southernmost state of Rio Grande do Sul), a small village near the shoe district of Novo Hamburgo. With close to 20 medium and large tanneries, pollution reached unbearable levels by the 1970s, and the district was one of the most polluted sites in the entire country. Pressure from a more than average environmentally aware local population and more stringent environmental regulations by the state authorities from the early 1980s forced the companies to start to clean up. In addition, UNIDO (United Nations Industrial Development Organisation) assisted the local authorities in the period between 1981 and 1987 to build pilot plants for treatment of tannery effluent in order to demonstrate practical solutions to the tanneries.

However, it was not until the early 1990s that increased national and international environmental awareness resulted in stricter environmental regulations that also influenced the tanneries outside the tanning districts in the south. With the 1992 Earth Summit in Brazil (Rio de Janeiro) there was increasing domestic debate on environmental issues, and environmental legislation was updated and political practice improved. Since that time environmental law at the federal level has become much stricter, also requiring regulations at state level to be no less strict. In 1991 a new comprehensive environmental law was presented to the Congress, but got stuck in the political bureaucracy. It was only in April 1998 that the law on environmental crime finally passed.

Even so there were improvements both in the environmental regulations and enforcement targeted at the tanneries during the 1990s. In the southern state of Rio Grande do Sul it was claimed that the stringency of the environmental law was the same as in the European Union (EU) and on some points even stricter. At the firm level there was a reasonably high level of consciousness on the matter among company directors, and smaller or larger efforts were under way in all companies to improve the environmental performance. Both national and foreign chemical companies claimed that solutions to pollution problems were the main driving force of innovation in the tanning industry, and that solutions had been found to most problems. An environmental expert said that 'the industry and government have [all] the knowledge [necessary to eliminate the environmental problems of the tanneries] – the rest is affordability'.

Environmental enforcement

While the law is one thing, enforcement is another. Many sources claimed that only the state of Rio Grande do Sul was at a European level when it came to environmental regulation. In this context it is important to note that many of the environmental problems in the south of Brazil have been solved, not by introducing cleaner production technology, but by outsourcing the pollution-intensive wet–blue production to the *bluseiros* in the west central and northern regions. An environmental expert said without hesitation that 'many [tanners in the south] do only finishing just to avoid environmental problems'. In other words, the most polluting part of the process has been interiorised to regions where the state government puts much more emphasis on attracting and building industry than enforcing environmental regulation. Several informants with knowledge of the area confirmed that environmental pressure from the authorities, environmental NGOs and communities in the interior is much lower, so the enforcement is very low. Even though there are some significant environmental advantages from relocating wet–blue production close to the raw material sources,[12] these are easily offset by lower standards in the enforcement of the regulations of the pollution intensive wet–blue process.

There are several indicators that the level of stringency in enforcement of environmental regulations in Brazil is substantially lower than in Europe. Tanners in the south of Brazil claimed that their environmental expenses were 0.5–2 per cent of the production costs, an environmental cost level that was confirmed by Abicouro[13] in their general investigation in 1998.

This is significantly lower than in Europe, where the level is 3–6 per cent. The average environmental cost in the tanneries in Brazil was found to be one-third of that of tanneries in Italy in absolute terms. Tables 6.2 and 6.3 provide details of the differences between Italy and Brazil in effluent treatment costs[14] by stages of production and as part of total processing costs.

In comparing processing costs, the significantly higher costs in Italy were those of labour, energy and effluent treatment costs, while costs of capital and chemicals were higher in Brazil. The right-hand column of Table 6.3 shows that the effluent treatment cost in Brazil was roughly 40 per cent of that in Italy, while the total processing cost itself was about 90 per cent. While the share of the effluent treatment cost in Italy was almost 10 per cent of the total processing cost, it was only 4 per cent in Brazil. The difference in the cost of treating effluents in Italy as compared to Brazil made up almost 60 per cent of the difference in the total processing costs. Observe also from Table 6.2 that almost all of the effluent treatment cost was in the wet–blue stage in both countries.

The significantly lower environmental costs in Brazil result from a number of factors such as cheaper energy (76 per cent of that in Italy), less expensive labour (61 per cent of that in Italy), scale economies in treating effluent, but not least the less stringent enforcement of environmental laws in Brazil creating a cost advantage for the Brazilian tanners as compared with Italian competitors.

One example of this is Rio Grande do Sul, commonly known to have the most stringent implementation of environmental regulations in Brazil. However, the cost of storing sludge and other solid waste in this state was said by several expert sources to be a fraction of that in Germany, which has the highest cost level in Europe. Even though state regulations have resulted in centralised storage of waste, the fundamental technical requirements to

Table 6.2 Effluent treatment cost[a] in Italy and Brazil in 1997 by stages of production

	In Italy		*In Brazil*		*Difference*	
Main stages of production	*Cost*	*Share*	*Cost*	*Share*	*Cost*	*Brazil vs. Italy*
Salted hides to wet–blue	7.61	89.5	2.86	87.2	4.75	37.6
Wet–blue to crust	0.87	10.3	0.41	12.5	0.46	47.1
Crust to finished	0.02	0.2	0.01	0.3	0.01	50.0
Total	8.50	100.0	3.28	100.0	5.22	38.8

Source: Abicouro (1997).

Note

a Effluent treatment cost (in US cents per sq. foot) includes both energy for running the treatment plants and the cost of sludge disposal.

Table 6.3 Effluent treatment cost as part of total processing costs in 1997

Cost in US cents per sq. foot	*Italy*	*Brazil*	*Difference B/I*	
Total processing cost[a]	90.59	81.61	8.98	90.1
Total effluent-treatment cost	8.50	3.28	5.22	38.8
Share of effluent treatment cost (%)	9.4	4.0	58.1	

Source: Abicouro (1997).
Note
a Costs of hides, taxes, administration and transport are not included.

fulfil the Brazilian legislation are not complied with at the storage facilities.[15] If the regulations were completed it would increase storage costs of tanning sludge five to ten times, according to a representative of the environmental authorities. In another southern state, Paraná, the environmental agency responsible for monitoring the tanning industry did not know where the tanneries disposed of their solid waste. There were no centralised storage facilities for the tanning sludge in the state, and the agency representatives assumed that the waste was disposed of in regular municipal waste facilities or other locations where it would be illegal to store it.[16]

Although there has been no general assessment of the environmental performance of all Brazilian tanneries, there have been several in-depth studies at the state level as part of German–Brazilian co-operation (e.g. Rodriguez, 1994). These studies have in general concluded that most companies in the Brazilian tanning industry need to undergo fundamental technological changes in order to meet the new environmental regulations. The investigations also show a great variation in the environmental performance of the companies. While some companies have state-of-the-art technology at all levels of production (one tannery even purified the water going into the factory), a number of tanneries, even in a southern state like Paraná, did not have primary treatment. Informants could also tell of extreme cases such as one tannery in the north of the country that supposedly let the tidal water clean the factory of waste.

According to a number of sources the general view is that the further north one goes in the country, the less strict are the regulations and enforcement. Even so, during a field trip in the southern state of Paraná, observations were made of open-pit filtering. This implies that wastewater with sludge is channelled into open-air pits from where the water filters down through the ground into the groundwater. Documentation showed that almost half of the tanneries in the region had this type of effluent treatment in 1990 (IAP, 1997). In the tanning district of Franca (in the state of São Paulo) the tanneries have built a centralised effluent treatment facility, but according to environment and tanning experts, the facility was undersized and worked poorly. While the newer *bluseiros* in the west central region of the country were required to build treatment facilities as an integrated part of the factory, informants with experience from the region say that the facilities are either insufficient, not operating or not working well. Often they are only turned on when the company expects foreign visitors or inspections of the factory. In these areas the environmental authorities are often poorly financed, understaffed and have little power. In addition, the administrative capacity of the Brazilian states has been reduced in recent years by cuts in the public budgets, thus limiting the capacity of the government to enforce new laws.

Environmental improvements

Despite the above observations, there are improvements in the environmental performance of the tanning industry in Brazil. Pollution intensity is down and awareness among actors in the industry is generally increasing. The latest law on environmental crime, passed in April 1998, calls for imprisonment of those who commit crimes against nature. The logic behind this law is that financial penalties for this sort of crime will not have the desired effect because the polluter will pass the cost on to the customers. Even before this law was passed, several tanneries were closed down permanently or temporarily by state governments because of their failure to comply with environmental regulations. The state government of Paraná set 1998 as the final year for the tanneries to comply with the existing regulations, otherwise they would be shut down. Only 12 of the 30 tanneries that existed in the state in 1990 remained in 1998. According to environmental experts and other tanners, several were closed due to problems with financing the construction and operation of treatment facilities. Investment costs for treatment facilities have traditionally been very high in Brazil owing to the geographical spread of the companies, the almost complete absence of municipal waste treatment facilities and the lack of will among the tanneries located close together to build common treatment facilities. All in all this makes it very difficult to obtain advantages of scale and to lower the treatment costs.

An interesting indication of the increasingly stringent environmental regulations for the tanneries in Brazil is how the companies evaluate the importance of the Brazilian environmental regulations on their competitiveness with foreign tanneries. Many tanners claimed it is a great problem that competitors in Asia (especially China) have substantially weaker environmental regulations and enforcement than themselves.[17] In combination with low wages, this was considered a significant problem for the competitiveness of the Brazilian shoe industry and, therefore, the tanners. Ironically, none of the Brazilian tanners mentioned the similar difference between Europe and Brazil, where Brazil has the advantage of more lenient environmental regulations and lower labour costs.

On the one hand the Brazilian tanning industry is being pushed to carry out investments in *process* technology to comply with the environmental requirements. In addition, Brazilian tanners have higher environmental standards and costs than tanneries from countries whose *products* they compete with, such as China. On the other hand there is no demand for 'cleaner' leather either in the domestic or the global market. As a consequence the tanning industry in Brazil is subject to an 'environmental squeeze', as investments in environmental improvements are a cost without economic return for the already heavily indebted tanneries. Similar pressure to improve environmental standards is exerted by the requirements put forward by governments in Europe, especially Germany, on imported leather. The ban against the use in leather of PCP (pentachlorophenol), certain biocides and a small number of azo dyes, has forced the Brazilian exporters to change their practice. At the same time the environmental investments are low and the local environment continues to suffer from contamination from the tanneries.

Since there is no price mechanism that can compensate for the increased environmental costs of the Brazilian tanners, and since enforcement is not always effective, the extent to which the tanners comply with the environmental regulations depends heavily on the attitudes of the company directors. As is to be expected there are great differences in these attitudes, related to the economic situation of the tanneries, their size, their markets and the general environmental practice in the region where they are located. Larger and visible tanneries with a sound financial situation and a brand name to protect usually see environmental costs as a natural part of the overall production costs, and have the intention of 'returning the water back to the waterways as clean as when it entered the factory'. Their attitude is that the environmental cost should have been included from the start for the tanners to operate with real costs and avoid sudden investment problems. They also consider the environmental costs a lesser problem, as other costs are more significant and more volatile, such as the price of hides and, therefore, much more of a worry. On the other hand, there are tanners with economic problems and smaller production volumes directed mainly at the internal market. These tanners are generally not willing to improve their environmental performance until the government threatens to shut the factories down.

Liberalisation and environmental practice – the link

To be able to evaluate the influence of liberalisation on the environmental performance of the tanning industry, a critical question remains. How has liberalisation influenced the implementation of the new environmental regulations? During the 1990s, liberalisation and financial stabilisation created a very difficult situation for many Brazilian tanneries as the weaknesses the industry developed in the 1980s were exposed. At the same time, the federal government has not seen any reason to take measures to help the industry, since the restructuring of production into larger groups and companies has resulted in a higher production volume and a significant positive trade balance.[18] This contrasts with the situation until the 1990s when Brazilian tanneries not only thrived in protected home markets, but also had limited environmental regulations compared to the once dominant northern European tanning industry that was subjected to increasingly stricter regulations and enforcement.

Before looking at the main negative consequences of liberalisation on the environment, it should be mentioned that liberalisation also has had some advantages for the environment. Increased exports to Europe resulted in certain environmental improvements in the Brazilian tanneries as they have been forced to comply with German requirements on the use of chemicals to preserve the leather for transport and storing. Since the production of wet–blue under the ruling conditions is profitable, there is a possibility that the *bluseiros* will be better able to make investments to comply with the environmental regulations, although this was not confirmed in this study. Another advantage of the liberalisation is that the tanneries, first of all those who can afford the capital costs, have better access to advanced technology both for improving product quality and environmental performance. Through better opportunities for partnerships with foreign companies, Brazilian producers can get both access to better technology and capital at lower interest rates.

Trade-off between industry and the environment

There are two main areas where economic liberalisation clearly has affected the environmental situation of Brazil negatively. Both have to do with the ways in which the local authorities and the companies deal with the difficulties arising from the changes in federal economic policy. Even if the production volume has grown, the number of companies and employment in the tanning industry has been more than halved. This has for many states resulted in reduced state-level tax earnings and employment, both as a result of less income tax and no export duty on wet–blue. Although the regressive restructuring is not necessarily caused directly by stricter environmental regulation, it is undoubtedly a situation that surviving companies can take advantage of. Pointing to their difficult situation they ask local authorities for time and support to fulfil the requirements, or otherwise they claim they will go bankrupt. Owners of tanneries who comply with the environmental regulations claimed these companies use environmental regulations as an excuse for their general poor performance.

No doubt this mechanism is part of the reason for the lower cost of treating effluents in Brazil. Although parts of the lower environmental cost can be attributed to scale advantages of treating the effluents, and lower labour and energy costs in Brazil, there are clear indications that it is also due to the lack of enforcement by the authorities. A number of informants explained that government enforcement of environmental regulation is reduced when the tanning industry or individual companies are economically fragile: the harder the economic times the less governments pressure companies to comply with environmental regulation. For Brazil and many other countries, when the political choice in hard times is between maintaining industry and employment on the one hand and protecting the environment on the other, the former wins out. This is all the more likely when environmental awareness among the voters is virtually non-existent and unemployment is high or growing.

Increased pollution through wet–blue

Perhaps the clearest and most serious consequence of liberalisation on the environment is not related to the enforcement of the laws, but a result of the significant growth in the production of wet–blue. As mentioned above, nearly all the pollution in tanning results from the production of wet–blue, while it yields relatively little value added. This means that Brazil is specialising in the lowest value added and most pollution intensive stage of tanning (see Table 6.4).

This has largely been influenced by the way Brazil has chosen to take advantage of the ample supply of domestic hides through trade barriers. Until 1990 the government limited exports of hides and leather through regulatory measures and reduced the cost of hides through increased supply for the domestic manufacturers of leather goods. This created a cost advantage close to 20 per cent for the domestic shoe industry compared to hide prices in Europe, and most leather was used domestically (Ballance *et al.*, 1992). Since the late 1980s, but especially since 1994, Brazilian tanners have increasingly competed by meeting the international demand for low-

Table 6.4 Leather exports from Brazil in 1997 (weight, value and pollution cost)

	Weight		*Value*			*Pollution cost*
Product	*Tonnes*	*Share*	*US$ million*	*Share*	*Per kg*	*Share (%)*
Salted hides	14.5	6.6	12.0	1.6	0.8	
Wet–blue	168.0	77.5	390.0	53.0	2.3	87.2
Crust	11.0	5.0	133.5	18.0	12.2	12.5
Finished	12.0	5.5	178.0	24.0	14.9	0.3
Total	207.0	95.7	719.3	97	Ave. 3.5	100

Sources: Abicouro (1997, 1998).

quality wet–blue hides. As mentioned above the growth in production and export of wet–blue has been one of the most significant structural changes in the industry.

One of the two Brazilian tanning federations, Abicouro, commissioned a study in 1998 to identify the crucial factors underlying the significant growth in wet–blue export (Abicouro, 1998). Most of the growth has gone to Italy, the world's largest producer of leather shoes, but also to other countries in Europe with significant tannery industries and to Mexico. The study concluded that the tariff of 7 per cent on Brazilian crust and finished leather entering the EU was the decisive factor. Since the Brazilian tax on export of wet–blue was eliminated in 1994, but not the tax on hides, the EU import tax made wet–blue the only Brazilian leather without duty, creating a powerful incentive to increase the trade of wet–blue from Brazil to Europe (see Table 6.5).

In addition to this situation, there were other factors strengthening the development. To specialise in wet–blue had become the easiest way for the national companies to survive the new economic regime in Brazil. First of all, producing wet–blue reduces the problem of lack of capital. Wet–blue has a short production cycle (4 days as compared to 20 to complete the whole process) and gives a quick return on investment. This is also the process which is most automated and allows for economies of scale. Since wet–blue can be used for a wider range of end-products than leather further down-stream, this kind of production also leaves the tanneries more flexible in relation to the markets, further reducing the risk.

Another reason for the growth of wet–blue is that it is a way to escape two of the main competitive problems of the Brazilian leather industry. One is the low quality of raw materials;[19] 90 per cent of the Brazilian hides are classified in the two lowest quality categories of five, and the quality is falling as the tanners cut chemical costs to meet the fierce competition. The other is the lack of skills among Brazilian tanners and shoemakers to make internationally competitive products with these low quality hides.[20] Low hide quality creates

Table 6.5 Cost of leather[a] produced in Italy with Brazilian raw material

Exported from Brazil	*Salted hides*	*Wet–blue*	*Crust*	*Finished*
Cost of processing in Brazil	0.0	18.0	48.7	81.6
Internal transport in Brazil	2.7	1.2	0.3	0.3
Export tax from Brazil	9.2	0.0	0.0	0.0
External transport from Brazil	6.0	2.6	0.9	0.9
Import tax to Italy	0.0	0.0	9.8	11.9
Cost of processing in Italy	90.6	65.6	32.3	0.0
Total	108.5	87.3	92.1	94.8

Source: Abicouro (1998).
Note
a US cents for each square foot.

several disadvantages. First, it reduces the value of the hides and end-products as it restricts the use of the hides to cheaper products. Second, for the low quality hides to become internationally competitive they require processing with knowledge-intensive finishing technologies to cover up the scarred surface and to strengthen the weak texture; the lower the quality of the raw hides the more skill, advanced chemicals and effort are required. The international market for leather buys low-quality Brazilian wet–blue, but has much less interest in more knowledge-intensive Brazilian leather and leather products. In this way Brazilian tanners find themselves in a *low quality trap*. Italian tanners and tanners from a few other European countries are, however, able to create internationally competitive products from the Brazilian hides utilising knowledge, skills and brand names which the Brazilian tanners do not possess. This explains why Italy, the world leader in tanning, is one of the main clients and the main investor recently in the Brazilian tanning industry. Italy buys almost a quarter of all leather exports from Brazil in value terms. However, this makes up one-third of Brazilian leather exports by weight. The higher share of weight than value indicates low levels of value added, suggesting that the Italians buy mostly wet–blue. Of the total export growth from Brazil since 1993, all in wet–blue, the Italians bought almost half. This means that Italy's demand for wet–blue is an important engine of production growth in Brazil.

Table 6.6 shows that hides and leather imported to Brazil have a much higher unit value than Brazilian exports. This is a clear indication of the low appreciation of Brazilian leather in the international market, as compared to the leather the high-end domestic shoe producers need to import to be competitive internationally.

A general assessment by a number of informants is that specialising in wet–blue export is not profitable if one has to treat the pollution according to the regulations. Even if some companies in Brazil produce at a very large scale, a frequent statement by environmental experts is that today's production is

Table 6.6 Unit cost of leather[a] imported to and exported from Brazil in 1997

Product	*Import value (US$ per Kg)*	*Export value (US$ per kg)*	*Excess of export value over import cost (%)*
Crust	20.5	12.2	68
Wet–blue	3.3	2.3	43
Wet–blue (small)	6.4	1.7	276
Finished	13.2	14.9	−11
Salted hides	1.0	0.8	25
Average (weighted)	10.5	3.5	

Source: Abicouro (1997).

Note

a Taxes and transport costs not included.

only possible owing to the weak and incomplete enforcement of the environmental regulations.

Even so, the tanning federation Abicouro still considers the combination of Brazilian and European tariffs, which makes wet–blue the only type of leather that is not taxed for export to Europe, to be the critical factor behind the increased wet–blue production in Brazil. In the opinion of Abicouro the best way to solve this problem is for the Brazilian government to introduce an export tax on wet–blue that makes it more expensive than Brazilian crust or finished leather. This export tax could even be lower than the European import tax on Brazilian crust and finished leather. Abicouro argues that the European tariffs today work in a protectionist manner. While the pollution burden is pushed to Brazil, Italy and other European importers of Brazilian wet–blue perform the cleaner processes with most value added. Considering the dominance of the Italian tanners in the world and within the EU, it is not impossible that this is one of the motivations behind the present tariffs. Italian tanners processed 55 million hides in 1997, and 50 million were imported. Having the wet–blue imported from Brazil greatly reduces the pollution in Italy, where there are already absolute shortages of water and space in two of the three central tanning districts (Arzignano and Santa Croce) (Gjerdåker, 1998). Outsourcing of wet–blue by the Italian tanners has been observed in other countries in the course of this research project, such as The Czech Republic, Poland and Argentina, and is probably widespread.

A factor that aggravates the increased environmental problem from the production of wet–blue in Brazil, is the process of 'interiorisation' of wet–blue tanneries to central areas with lower environmental standards. The process of relocalisation from the south to the west central and northern regions of Brazil started simultaneously with the introduction of the first environmental regulations. Even so, the main motivation for the 'interiorisation' of wet–blue production was the advantage of being near to the relocated cattle herds and the slaughterhouses. In recent years, as Brazilian industry in general has been facing tougher international competition, 'interiorisation' has taken place to cut costs. When asking how important the less stringent environmental enforcement in these areas has been for motivating the continued 'interiorisation' of tanneries, the answers varies from 'crucial' (environmental experts) to 'insignificant' (tanners). Even if it is not the prime motivating factor, the lower environmental costs in the interior cannot be ruled out as one of several economic incentives for the tanneries to move to the interior.

No matter how strong the enforcement of environmental regulations in Brazil, the main fact remains that the country gets an increasing part of the pollution burden, and a smaller part of the value added. As wet–blue is exported, water and land in Brazil is contaminated without any significant cost to the consumers. The growth of wet–blue production is, and will be, generating significant profits for the industry as long as the industry can operate under the present conditions and demand in the market is good. In a longer time perspective the consequences of a Brazilian tanning industry dominated by exports of wet–blue is most likely to be detrimental to its competitiveness. If the quality of wet–blue continues to decline, the reputation of Brazilian hides will worsen, making it more difficult to change the attitudes of more advanced users in the domestic and international market if improved at a later date. A growing number of Brazilian shoe companies buy Argentinean crust for their production of more advanced shoes for the US market. This development can result in even weaker network relationships between domestic tanneries and shoe producers, reducing the chances of creating an internal innovative force in the industry. With increased reliance on wet-blue production, the need to maintain and improve skills to increase the value added is reduced, and the knowledge level and innovative capability of the Brazilian tanning industry may be further reduced. Statistics show that even the tanneries in the south, which traditionally are knowledge intensive and make finished leather, are producing more wet–blue. Yet another indication is that leading multinational chemical companies serving the tanning industry are drastically reducing their activities in Brazil, leaving the market to domestic firms. These are all perils of the low quality trap.

Final considerations

This chapter has shown that the Brazilian tanning industry has gone through a dramatic restructuring during the last decade. This has coincided with the liberalisation of the Brazilian economy and the introduction of environmental regulation in the country. However, the economic liberalisation and the environmental regulations are not the direct causes of the problems many of the tanneries have today. During the 1980s the Brazilian tanning industry operated within a set of conditions that created latent problems, whose consequences became clear only after the industry was exposed to global competition. Problems such as the *low quality trap* and the *environmental squeeze* are expressions of the Brazilian tanners' limited preparedness to handle increasing international competition and environmental regulations. To survive, the companies have to an increasing degree turned their activities towards an area where they are most competitive within the new regime, specialising in low quality, low value added and pollution-intensive production of wet–blue for export.

At first sight this seems to be something that the Brazilian authorities should worry about. The federal government is, in line with its economic stabilisation plan, first of all interested in achieving a national trade surplus and economic growth, and increasing international investments. The state governments are more concerned about the survival and establishment of companies, such as tanneries and the level of value added, tax earnings, employment and the environment. In the insecure economic situation for domestic companies, the state governments are prone to give priority to the survival and establishment of industry, even when this is at the expense of the environment.

Some will argue that the changes in the tanning industry in Brazil are a necessary and healthy adaptation to the realities of the international economy. On the other hand there are consequences for employment, value added and the environment. While the intention of those who push for economic liberalisation might not be to create a counter-force to environmental improvements, the consequence of this economic policy is a clear negative impact on the relative environmental performance of the tanning industry in Brazil. It would not be a surprise to this author if further investigation revealed that these mechanisms had the same consequences for other industries in Brazil and other countries. There is undoubtedly a need to find ways to ensure that economic liberalisation does not further increase the pressure on the environment.

Notes

1 The empirical basis of this paper is nearly two years of study of the leather industry as part of the European Union funded project *Environmental Regulation, Globalisation of Production and Technological Change* (ENV4-CT96-0235), of which the Department of Human Geography at the University of Oslo was one partner, and specifically a month of fieldwork carried out in Brazil in April/May 1998. Directors in 12 companies in the leather industry and 15 highly qualified tanning experts were interviewed on the Brazilian tanning industry. Due to an agreement with the informants of sending them direct quotes for approval before using them in presentations, the individual sources will not be mentioned by name in this paper.

2 I am grateful for comments by Hege Knutsen, Anne Gjerdåker, Jan Tore Odegard and UNIDO's leather section.

3 In this chapter the term environmental regulation refers both to legislation and its enforcement.

4 'Wet–blue' is a term referring to semi-processed leather obtained from hides or skins through the chrome tanning process, after which the colour of the tanned leather is blue–grey. The denomination 'wet–blue' is given as the leather tanned with chrome is kept in a wet condition, and it is also traded as such (covered in plastic). 'Crust leather' is leather which after wet-blueing is passed through the additional processes: splitting, retanning, dyeing, fat-liquoring and drying. 'Finished leather' is obtained by the final processes given to crust leather to make it ready for use in the manufacture of different types of leather products, such as buffing on the grain side, application of pigments and other finishing chemicals through roller-coating or spraying, as well as final plating, to give a smooth or patterned leather surface. (All definitions provided by UNIDO's leather section.)

5 Light leather has been primarily tanned with chromium, while heavy leather is primarily tanned with vegetable agents.

6 *Regressive restructuring* is a term for the restructuring of an industrial sector, leading to clear negative development for the companies in the industry, in regard to production volume, number of companies and skill level.
7 This is a rough estimate as the increase of value per kilo varies greatly with type and origin of the raw material, and also varies with time of purchase.
8 Raw hides makes up between 50 and 60 per cent of the production costs of finished leather.
9 The growth in Brazil's shoe industry in the 1950s was the prime engine of the development of the tanning industry. In the mid-1980s, 80 per cent of total production was shoes for the domestic market and 70 per cent of export was to the USA. Cheap labour, abundance of raw materials, active state intervention and a reasonable level of industrialisation were crucial factors for creating a shoe industry with 300,000 employees, made up of between 2000 and 4000 companies according to different estimates (depending on the inclusion of artisan cottage industries).
10 Brazilian shoe producers had 34 per cent of the US leather shoe market in 1985, and 30 per cent in 1990. In 1993 Brazil exported more than 130 million pairs of leather shoes to the USA, but this declined to 96 million pairs by 1995. At the same time China increased its export of leather shoes to the USA from 1 per cent of the market in 1985 to 15 per cent in 1991 (425 million pairs), and further increased the sales to 716 million pairs in 1995 (US Leather Industries Statistics, 1997).
11 BOD/COD: biological/chemical oxygen demand, or the amount of oxygen required in the water to break down the organic or chemical material contained therein. High levels of BOD or COD are a threat to life in waterways that is also dependent on oxygen.
12 The advantages are that the use of salts and chemical preservatives for the preservation of hides for transport is almost eliminated, and less energy is used for transport since wet–blue weighs half as much as raw hides.
13 Abicouro is a new industry federation formed in 1996 by 70 tanners mainly in Rio Grande do Sul and São Paulo. It is dominated by high-value-added companies that see the growth of the wet–blue exports as detrimental to the industry.
14 Effluent treatment costs represent the bulk of environmental costs for the tanneries.
15 Since most tanneries use chromium in the tanning process, which results in high amounts of this heavy metal in the tanning sludge, the solid residues are classified as special waste and must be disposed of in specialised locations (hence the high cost).
16 Since sludge with chromium is defined as specialised waste, it can according to the law only be stored in special storage sites.
17 To what extent the Brazilian tanners are correct in their evaluation of the environmental performance of the Asian, and especially Chinese, tanneries is a matter of doubt. There are nationwide regulations and enforcing authorities in China, which, as an example, resulted in the closing down of several hundred tanneries in 1997 and 1998 (*Leather Manufacturer*, 1998).
18 In 1997 the trade balance surplus for leather was US$ 570 million, and total leather exports (including shoes and artefacts) accounted for 1.4 percent of all Brazilian exports (Abicouro, 1998).

19 The low quality of the Brazilian hides is due to a number of factors: breeds with thin hides, warm climate, extensive cattle raising, low awareness and priority to hide quality among cattle owners and slaughterhouses. Despite efforts in recent years to improve the quality of the raw hides, the general opinion in the leather industry is that there has been a marked reduction in the quality of Brazilian wet-blue.

20 The reason that the tanning know-how in Brazil is not advanced enough to provide internationally competitive leather from domestic hides is due to Brazil being virtually closed to outside competition until the mid-1980s. This monopoly situation created a culture of little competition and high profits, where most of the income was taken out of the company, and little capital saved or reinvested in quality/environmental upgrading or improving the skills of the coming generations. Much capital was diverted to financial speculation. With increased competition in price, diversity, quality and flexibility following the liberalisation, the mismanagement of many companies was exposed, and the investment burden to catch up technologically became too high for many indebted tanners. Soaring interest rates limited access to capital for domestic companies, but also weakened the client industries of the tanners. Furthermore, the structure of the industry was not conducive to learning through user–producer relations either between the companies, or with the users (shoe producers), owing to the comfortable situation of the all-absorbing domestic market of the 1980s. Brazilian products were (and still are) copies of last year's fashion in Europe, and the internal dynamic was weak. Many of the shoe producers catering to more advanced export markets became frustrated by the inflexibility of the established tanners, and helped start up new tanners that would cater to their special needs. This forced the traditional tanners to improve their co-operation with shoe-producers to meet international quality requirements.

Bibliography

Abicouro (1997) *Boletim Estatístico do couro 1997*. Novo Hamburgo: Abicouro.

Abicouro (1998) *Síntese do Estudo da Competitividade dos Curtumes Europeus e Brasileiros/ Argentinos*. Novo Hamburgo: Abicouro.

ABQTIC (1993–7) *Guia Brasileiro de Couro*. Estancia Velha.

Ballance, R.H., G. Robyn and H. Forstner (1993) *The World Leather and Leather Products Industry – A Study of Production, Trade Patterns and Future Trends*. Liverpool: UNIDO/ Shoe Trades Publishing Limited,

FAO (1996) *World Statistical Compendium for Raw Hides and Skins, Leather and Leather Footwear 1977–1995*. Rome: Food and Agriculture Organization.

FAO (1997) *World Statistical Compendium for Raw Hides and Skins, Leather and Leather Footwear 1977–1996*. Rome: Food and Agriculture Organization.

Gazeta Mercantil (1998) *Balanço Anual 1997*. São Paulo.

Gjerdåker, A., 1998. *Miljøreguleringer og konkurranseevne i garveri-industrien: Et eksempel fra Santa Croce sull'Arno*. M.Phil. thesis in Human Geography, University of Oslo, Norway.

IAP (1997) *Manual de Controle de Polução Hídrica de Curtumes – Situação dos Curtumes Implantados no Estado do Paraná*. Curitiba: Instituto Ambiental do Paraná.

Leather Manufacturer (1998) Arlington, MA: Shoe Trades Publishing, Vol. 116 (various issues).

Miljøministeriet (1992) *Branchevejledning for forurenende garverigrunde.* Copenhagen: Vejledning fra Miljøstyrelsen, No. 5, Luna-Tryk Aps.

Rodriguez, A. (1994) *Diagnóstico sobre curtumes do norte, noroeste, oeste e sudeste do Estado do Paraná.* Curitiba: Centro de Integrações e Tecologicas do Paraná (Projeto de Cooperação Técnica Brazil – Alemanha – GTZ/ABC).

UNIDO (1997) *Mass Balance in Leather Processing.* Vienna: UNIDO.

US Leather Industry Statistics (1997) US Department of Commerce (www.ita.doc.gov/td/ocg/leather.htm).

World Leather (1997–8) London: Shoe Trade Publishing, Dec.–Jan.

7

INDUSTRIAL INNOVATION AND ENVIRONMENT IN THE PULP EXPORT INDUSTRY IN BRAZIL[1]

Sonia Maria Dalcomuni

The renewal of environmental consciousness in society from the mid-1980s, together with the corresponding growth in the extent and stringency of related regulations, have brought about a change in the nature of the environmental pressures on the pulp and paper industry because these issues began to affect the marketplace directly. By becoming market-related factors, such pressures have promoted an associated change in the nature, source, means and geographical scope of firms' environmental responses in that industry. As a result, important changes in their behaviour regarding environmental issues have been seen in the past decade: from reactive to proactive, and from isolated and peripheral add-on measures absorbed by firms from suppliers to a global trend towards intertwining firms' environmental responses with their core technological and marketing investments and decisions.

This chapter aims to demonstrate on the one hand that recent developments in environmental regulation in Europe of eco-labelling and environmental management systems have had worldwide effects on innovative activities in natural resource based industries by firms for whom European markets are important destinations for their products. In such markets, to be and to seem to be green became essential – in order to be competitive. On the other hand it is also stressed that, although markets can signal the changes needed in environmental innovation, they do not automatically provide the means for so doing. Firms' abilities to cope with such a changing environment depends especially on their internal production capabilities in promoting green innovation, namely, the dynamic technological, managerial, financial and marketing capabilities to survive and succeed in facing such challenges. Thus, in order to foster 'green innovation' within firms, there has to be greater focus on the analysis and design of regulation within firms, where innovation actually takes place.

For that purpose research was carried out from 1995 to 1997 involving interviews with both 51 professionals within the five large pulp export firms in

Brazil (Aracruz Celulose SA, Cenibra SA, Riocell SA, BahiaSul SA and Jari Celulose SA) and 34 professionals from regulatory agencies, firm associations, research institutes and finance agencies in Brazil and in Europe. The research focused on those firms' environmental experiences. It highlighted the widespread effects of environmental issues on industry innovation and economic practices as a possible important component of the trends towards globalisation. It also revealed that, in order to face the environmental challenge of putting industrial production on a sustainable track, it is necessary to develop an appropriate institutional apparatus to promote creation of capabilities and continuous improvement in environmental related techniques and procedures within firms.

Pulp and paper industry background

Pulp and paper constitutes a highly globalised, competitive and diverse industry worldwide. It accounts for about 2.5 per cent of the world's industrial production (UNIDO, 1993; IIED, 1996). This production takes place in about 100 different countries in all five continents. Paper production provides a wide range of paper grades aimed at meeting an even wider range of consumption needs, including communication, packaging, household and sanitary, among others.

International trade in the pulp and paper industry comprises sales of wood chips, market pulp and paper. It is a growing industry. Global annual consumption of paper is about 270 million tonnes. It has increased more than five-fold in the period from 1950 to 1991. There are also expected to be continuous paper consumption increases over the next two decades. The FAO (1995) has projected that the demand for pulp and paper products will continue to grow, and should more than double in developing countries by 2010.

The world's pulp and paper production and consumption are highly concentrated in geographical terms. Approximately 90 per cent of the total output of paper in 1996 was consumed in a combination of North America (35 per cent), Europe (25 per cent) and Asian Pacific (29 per cent). World pulp and paper production is illustrated in Table 7.1.

According to the degree of vertical integration in the pulp and paper industry, firms can be characterised as follows:

1 *integrated firms*: firms that focus on paper as the main product and encompass the production of wood and pulp;
2 *non-integrated firms*: firms that focus on paper as the main product but depend on third parties for supply of pulp;
3 *market pulp firms*: firms that focus on pulp as the main product, for third party supplies for papermaking.

Table 7.1 World's pulp and paper production

Region	*Paper (%)*	*Pulp (%)*
North America	36	40
Europe	33	29
Asia	24	23
Latin America	5	6
Oceania	1	1
Africa	1	1

Source: IIED (1996).

The characteristics of high capital intensity and large-scale production of the world's pulp and paper industry has been stressed by various publications on this industry (BNDES, 1996, pp. 47–72). It has also been suggested that concentration in this industry is still increasing.

The pulp and paper industry has historically aimed at upstream vertical integration in order to guarantee access to virgin fibre. The traditional highly verticalised production suggests a high degree of self-sufficiency of fibre input in the main paper-producing countries. In fact, about 75 per cent of the pulp produced worldwide is consumed in integrated paper mills. However, a substantial trade in market pulp has developed quite rapidly in recent decades. In 1993 about 20 per cent of world pulp production was sold as market pulp.

Canada, the USA and Scandinavia have traditionally been the major producers of market pulp, accounting for more than 80 per cent of the total output during the 1960s and 1970s. However, the entry of new suppliers from countries such as Brazil, Chile, New Zealand and Indonesia has eroded the northern producers' market share to about 65 per cent today (IIED, 1996). North America remains the biggest pulp and paper producer, consumer and net exporter region in the world. However, in terms of market pulp production, country positions have changed considerably in the period. Most of the new entrants have relied for their market pulp production on fast-growing trees genetically modified through biotechnology. Figures 7.1 and 7.2 illustrate the market pulp patterns.

The pulp and paper industry in Brazil is made up of two quite distinct industrial segments. In the paper industry there is a predominance of family-based economic groups, even among firms that became large and are leaders in the industry. Historically, the sales of the paper segment have been oriented to the domestic market. There is much greater heterogeneity in the paper industry as regards firm sizes, managerial and technological capabilities and degrees of verticalisation. The market pulp segment in Brazil consists of five firms: Riocell SA, Jari Celulose SA, Cenibra SA, Aracruz Celulose SA and BahiaSul SA. These firms were established mainly through international joint ventures and direct foreign investments. Their clients are

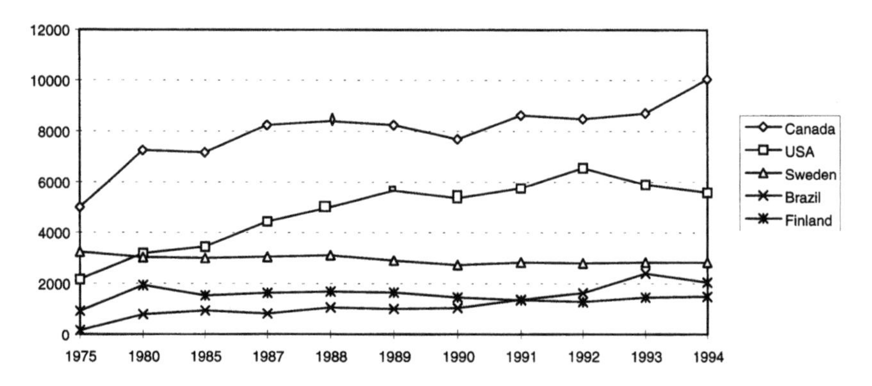

Figure 7.1 Five largest pulp export countries, 1975–94 (000 tonnes). Source: PPI (1996) *International Fact and Price Book.*

mainly paper makers located in OECD countries (especially Europe, the USA and Japan) (Firms' Annual Report, various issues). They are large scale and vertically integrated industrial plants.

The market pulp industry was established in Brazil as a result of the dynamic interactions of such factors as: the economic opportunities made possible by both the national and the international economy from the mid-1960s; the public policies relating to industry finance, science and technology and forestry; the entrepreneurial initiatives of foreign investors and of entrepreneurs in activities other than paper production; the development and continuous improvement of the technology for the use of eucalyptus for paper production; and finally the learning processes in wood production and paper-making activities that took place both within firms and in research institutions in the country (BNDES, 1993a,b; Dalcomuni, 1990).[2]

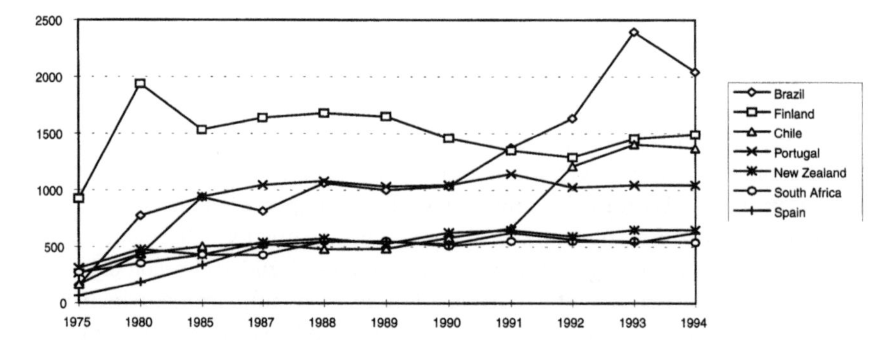

Figure 7.2 Market export patterns of selected countries, 1975–94 (000 tonnes). Source: PPI (1996) *International Fact and Price Book.*

The market pulp export industry began to be established in Brazil during the mid-1960s with start-ups concentrated during the 1980s. The Brazilian state played a major role in this process. In the 1970s foreign firms (Jari and Riocell) began to be totally controlled by the state. Reprivatisation began in the mid-1980s, a process somewhat intensified during the 1990s due to the privatisation of CVRD (Companhia Vale do Rio Doce) a large mining company and an important shareholder in two of the five Brazilian market pulp export mills. However, as a recent phenomenon no significant change in firms' performance has been observed up to now as a direct result of privatisation.

Production process and environmental impacts

The pulp industry is considered to be the third most polluting activity worldwide (after energy generation and the chemical industry). The pulp and paper production chain encompasses wood, pulp and paper production. The environmental issues relating to these activities are various. In wood production, the main focus has been on land use and ownership, biodiversity and forestry management. In the pulp industry the focus has been on liquid effluent emissions, air emissions, solid waste disposal and the toxicity of final products. In the paper industry, the focus has been on the toxicity of final products and emissions, including paper recycling and solid-waste disposal issues.

The general principle of all pulping processes is the same: they are designed to separate the fibres in the wood from the lignin, a natural glue that connects the fibres. The more the lignin can be removed without damaging the fibres, the better the quality of the product in terms of strength and bleachability requirements. Figure 7.3 illustrates a chemical pulping process.

With regard to pulping processes for virgin fibres there are, broadly speaking, two main types: chemical (42 per cent of the total) and mechanical pulping (15 per cent), together with a combination of both which results in semi-chemical pulping processes. Among chemical pulping processes, sulphite and sulphate or Kraft processes are the main ones in use. Completing the main pulping processes used in the world pulp industry at present are recycled fibres (26 per cent) and de-inking pulp (8 per cent) for non-virgin fibres.

The main environmental impacts of pulp production can be summarised as follows:

1 *Liquid effluents*: emissions to water from the pulp and paper industry have been a major focus of public concern. They vary according to the type of pulp and paper production and include dissolved wood substances, residual chemicals and compounds produced by reactions between wood substances and residual chemicals. Liquid effluent emissions are generated from the

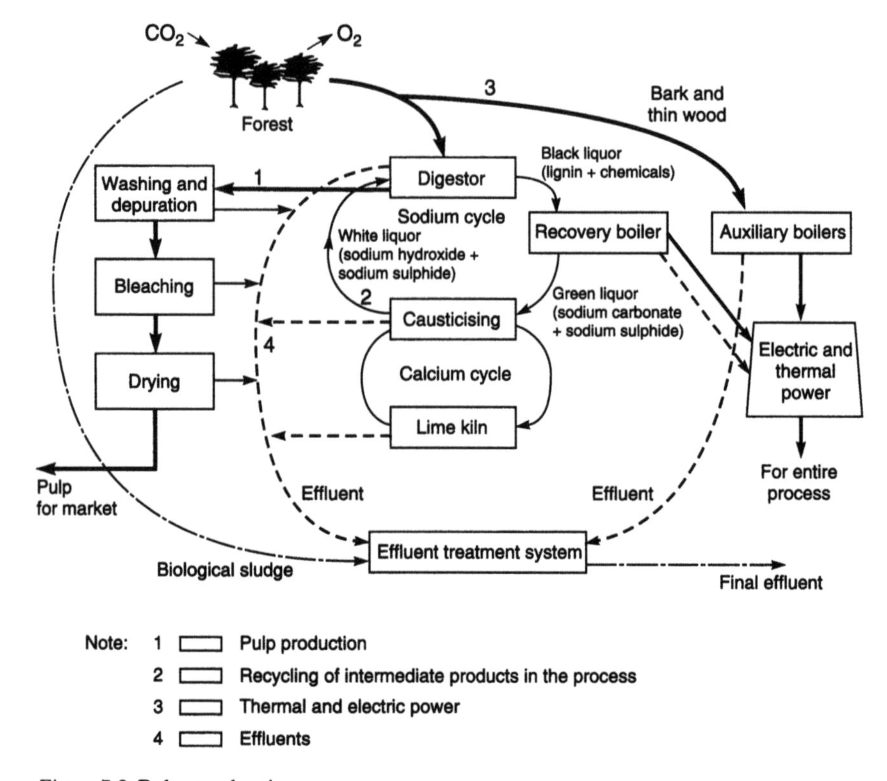

Figure 7.3 Pulp production process.

bleaching and washing phases, from condensates spills, screen rooms and wood handling. These emissions are analysed through the main groups: BOD (biological oxygen demand), COD (chemical oxygen demand) and suspended solids.

2 *Air emissions*: air emissions from the pulp and paper industry include sulphur dioxide (SO_2), nitrogen dioxide (NO_2), hydrogen sulphide (H_2S) and solid particulates (dust). Air emissions have been one of the main sources of environmental pressures on chemical pulp mills, resulting from the emission of malodorous gases of reduced sulphur compounds. These gases are considered harmless; however, their foul smell tends to cause local nuisances. These emissions are mainly generated by boilers.

3 *Solid waste*: solid wastes generated by the pulp and paper industry include: bark, wood residuals, mineral ash, dregs/grits and sludge among other less important ones. This type of pollution normally receives less attention in the technical publications addressing pollution aspects in this industry. However, recycling and solid waste disposal have grown in importance in the environmental debate in OECD countries.

The pulp industry has always been the subject of environmental pressures and pollution abatement investments. However, from the mid-1980s onwards, these pressures intensified. The bulk of the innovative effort in production processes for the pulp and paper industry as a whole was concentrated in market pulp. The rate of environmental innovation and the speed of its diffusion have also increased substantially. At the same time, regulation on this industry has become more complex and wider in scope, mainly in Europe.

Since 1985, the industry has been especially targeted for environmental pressures. The pulping phase, considered the most polluting one in the paper production cycle, has been the epicentre of these pressures. The landmark that prompted the massive environmental pressures directed at the pulp industry during the past decade was the discovery of polychlorinated dioxins and furans in bleaching plants. Dioxins are alleged to have carcinogenic links, and the elemental chlorine used in bleaching was quickly identified as a source of potential toxicity. In this context environmental pressure groups campaigned for phasing out the use of chlorine in the industry. At the same time, many government regulations in the major industrialised countries were drawn up and implemented, with particular concern about the generation of dioxins, not only in the pulp industry. The focus of such regulation is on reducing levels of the parameter group AOX.

As a result there has been innovation and diffusion of cleaner technologies for alternative pulp-bleaching phases in the pulp industry in various countries (Collins, 1994; IIED, 1996). This diffusion process, if compared to the historical technological patterns in the pulp and paper industry can be considered to have been relatively fast. Thus, ECF (elemental chlorine free) and TCF (total chlorine free) pulping processes were developed, which aimed to substitute or phase out the use of chlorine in the pulping processes. In addition, research and pioneering attempts at completely closing the loop of pulp and paper production have also been the subject of scientific scrutiny in many firms worldwide (TEF – total effluent free).

Environment-related investments in the sector have not been limited to innovations in the pulping phase. The past decade is also characterised by an intensification in investments in pollution abatement in developed countries in recent years, reinforcing a historical process of cleaning up the industry, which began mainly from the mid-1960s. This pattern is illustrated in Fig. 7.4.

Trends in environmental awareness and performance in the world pulp and paper industry

Major awareness and trends related to environmental issues observed in the world pulp industry in the past decade are summarised as follows:

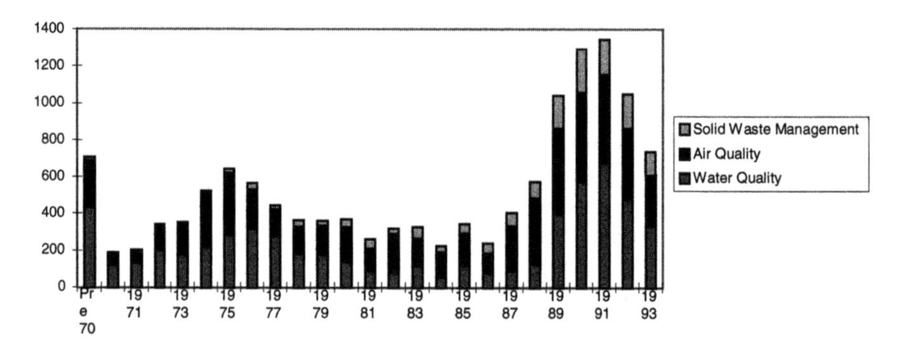

Figure 7.4 Environmental expenditures ($m) for pulp and paper industry in the USA, pre-1970–93. Source: *North American Fact Book* (1994).

1 Awareness concentrated on liquid effluent emissions – possible dioxin generation by the use of chlorine in bleaching phases.
2 Intensification of clean-up investments in countries such as Canada, the USA, Scandinavia and central Europe.
3 Increasing number and sophistication in terms of design and execution of environmental regulation related to the pulp industry.
4 Trend towards process innovation and 'green' product development in some leading pulp-producing countries with the development of ECF and TCF pulp to replace standard pulp.[3] This trend worldwide is illustrated in Table 7.2.
5 An intensification of green consumerism in Europe.

Development of environmental regulation in Europe and North America gave particular attention to the pulp industry. Initially led by the debate on the 'chlorine issue', environmental regulations not only grew in number and stringency, but also began to target 'green consumerism' more clearly, as previously mentioned. Hence, pulp firms have been especially 'pushed' towards developing cleaner production processes and products. Investments in environmental control in this industry are nothing new. Because of the polluting nature of the pulp industry, it has long figured as a special target for regulation and pressures from environmental campaigners. As a result, firms have invested in cleaner production to comply with legislation, such as environmental standards and emission limits, and to meet market demands or seek opportunities in new market niches. Table 7.2 illustrates the worldwide diffusion of cleaner production techniques.

In terms of the design of environmental regulations there has been a tendency to shift away from linear and rather punitive command-and-control approaches to more participatory and negotiated agreements. Europe has played a leading role in such a process.

Table 7.2 Regional distribution of bleaching techniques

Continent	*Unbleached (%)*	*Standard (%)*	*ECF (%)*	*TCF (%)*
Central Europe	22	20	33	25
Scandinavia	25	5	55	15
Oceania	38	39	10	13
Southern Europe	18	38	36	8
Canada east	14	61	23	2
Brazil	25	50	23	2
Eastern Europe	32	50	16	2
United States north	7	76	15	2
Canada west	9	41	50	–
Indonesia	7	70	23	–
Other Latin America	32	48	20	–
United States south	42	46	12	–
United States west	47	50	3	–
Russia	56	42	2	–
Asia BMP	–	100	–	–
Asia KTT	18	82	–	–
Japan	18	82	–	–
India	19	81	–	–
South Africa	28	72	–	–
Other Africa	57	53	–	–
China	77	23	–	–

Sources: Jaakko Pöyry (1996); IIED (1996). Data are estimates based on a chart from these studies.

Notes

Asia BMP: Bangladesh, Malaysia and Pakistan.

Asia KTT: Korea, Thailand and Turkey.

According to Price (1994) there were some 200 pieces of European Union (EU) legislation concerning the environment in force in 1994, with 50 waiting to be adopted and another 50 at the conceptual stage. Recent EU environmental legislation has introduced quite a few new principles and procedures with a direct impact on the pulp and paper industry. These include targeting changes in consumer's behaviour (1992); legislation on access to information; legislation on remedying environmental damage; the packaging waste directive; legislation on eco-management and auditing systems (EMAS, 1993); eco-labelling systems (1992); timber certification; and proposals for new taxes (carbon/energy tax; the eco tax and a pulp tax). In addition, in the above-mentioned countries, there has been increasing stringency on emission standards' setting and adoption, and a trend towards banning the use of chlorine.

Initially prompted by the debate on the 'chlorine issue', environmental regulations not only grew in number and stringency, but also began to target 'green consumerism' more clearly. This multiplication and increasing stringency in environmental legislation explains, in part, the pulp firms' intensification of environment-related investments and research efforts to develop cleaner production processes. Therefore, regulation figures as a first reason for environmental-related investments by pulp firms.

Another reason for intensifying such investments is that their markets began to demand cleaner products and production practices from the late 1980s onwards. The origins and intensification of such 'green consumerism' in the pulp industry has been geographically concentrated in Europe. From a direct necessity to meet customer demands for 'greener products', investments in cleaner production have been carried out in order to exploit new market niches, e.g. the development of TCF pulp initially to supply German markets. The main driving force for these changes in the production process was the search to replace or eliminate the use of chlorine in the pulping process, in response to environmental concern and pressures.

The spread of actual and potential 'green consumerism' in Europe has had as its main driving force the development of market-based mechanisms in European environmental regulation, especially the EU eco-labelling scheme. Despite its basis in already existing national schemes in various OECD countries, setting up the EU eco-labelling system has promoted more intensive debate and more extensive actual effects on the environment-related investments of pulp firms than any of its predecessors. More importantly, the historical local or national effects observed in complying with environmental legislation have been changed as effects have spread far beyond the political administrative boundaries for which they were initially defined, including their influences on the market pulp export firms in Brazil.

Of greatest importance have been changes in most ABECEL firms' customers in Europe towards products with less or no content of chlorinated organic compounds. From an initial focus on products, such demands soon began to include requirements for cleaner production practices in pulp mills, with special attention to the content of chlorinated compounds in liquid effluent emissions. These demands have been continuously added to, for example, introducing sustainable management practices in forestry and increasing the amount of recycled fibre content in paper.

By transforming environmental aspects of pulp and paper products and their production processes into an additional consideration that customers have to take into account in deciding about their purchases, eco-labelling schemes have made environmental concerns endogenous to firms' economic and technological dynamics. In this manner, the environment has taken its place on the research agenda for their R&D.

In so doing, 'green innovation' has also been integrated into the innovative dynamics of pulp firms.[4] These competencies have become an important basis for their survival and growth in European pulp markets. Concerns about continuous improvement in environmental performance were observed to be more evident in firms that also appeared to be keener and more able to implement continuous improvement in their general economic and technological activities.

The research also showed that, of equal importance as strength in technological capabilities within firms, successful 'green innovation' is dependent on focusing R&D on the environment. In addition, it was also observed that firms' capabilities to innovate – including green innovation – are highly stimulated or constrained by the national economic and institutional framework within which they are located. The combination of firms' internal innovative capabilities with strategic partnerships established for innovative purposes ultimately define their success or failure in this respect.

Environmental performance in the Brazilian pulp industry

The research carried out within Brazilian pulp export firms revealed that environment-related investments and research effort within them followed a similar path to that observed in the pulp industry internationally. Firms promoted process changes, product changes, organisational changes, and an intensification of environment-related research. Customer services' development and reinforcement of inter-firm co-operation for discussion, benchmarking and lobbying purposes were also observed as intensifying in the past 10 years. All of them demonstrated an understanding that linking environmental issues with economic performance and strategies is a permanent structural change in this industry. Finally, data provided by the firms illustrated that, despite having invested in environmental protection since well before 1985, during the last decade all these firms have intensified their initiatives to tackle cleaning-up and process-integrated measures.

Brazilian pulp export firms have focused their green innovative responses on three main issues: development of new bleaching processes, adoption of environmental management systems (ISO 14000) and monitoring of environmental regulation in the European Community, especially the eco-labelling schemes. Adoption of cleaning-up measures actually took place long before 1985, but, in the last decade, data on emissions provided by firms exhibit significant reductions in pollution generation by their industrial activities, as illustrated by Fig. 7.5 on liquid effluent emissions at Aracruz.

The analysis of the timing of such responses revealed a clear shift in the nature of firms' environmental responses from the late 1980s onwards. Investments in cleaning-up measures represented the historical focus of their attention and were mainly driven by compliance with local regulation and acceptance by local communities. This resulted in highly firm-specific experiences and related environmental performance in Brazil and abroad. Recently, cleaning-up measures, especially those related to liquid effluent,

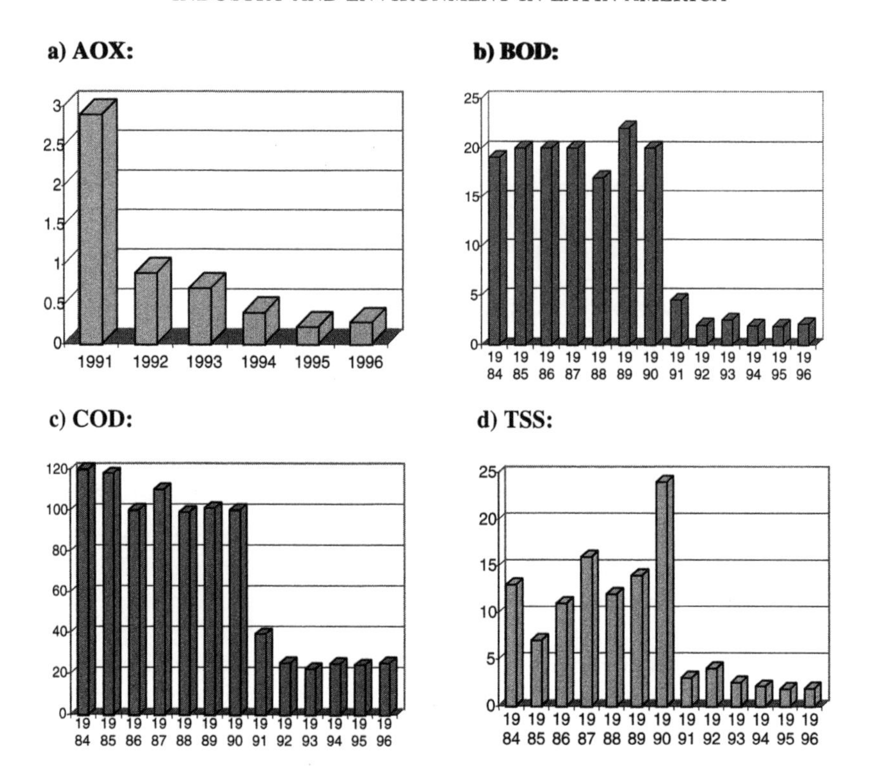

Figure 7.5 Discharges to water at Aracruz, 1984–96 (kg/tonne). Source: Aracruz Celulose SA.

have been tackled using a process-integrated approach. New research on effluent toxicity was also observed as intensifying and spreading among firms. The main driving force behind such a change has been the controversial 'chlorine issue'.

Despite this recent renewed focus of concern on 'the chlorine issue', with regard to effluent treatment and solid residual treatment or disposal, Brazilian pulp export firms have historically presented an average to above-average performance when compared to their counterparts in OECD countries. This is illustrated by Table 7.3 on the ratings of discharges to water defined internationally. With regard to solid-waste issues the experience of Riocell, of implementation of a solid waste-free mill project launched during the 1980s, enabled that firm to recycle about 98 per cent of the waste generated in its production processes. Regarding liquid effluents and gases it is important to point out that BahiaSul presents outstanding low emission values compared to its counterparts in Brazil and abroad.

The most visible innovation/diffusion of cleaner production technologies in Brazilian pulp export firms, following an international trend in the leading firms of the world pulp industry, has been the development of ECF and

Table 7.3 ABECEL firms' liquid effluent emissions and ratings (kg/ADT), 1996

Firms	*BOD*	*COD*	*TSS*	*AOX*
Riocell	0.25 {1}	3 {1}	1 {1}	< 0.2 {1}
Aracruz	2.3 {1}	26.9 {2}	2.4 {1}	0.3 {1}
Cenibra	3.8 {1}	18.0 {1}	2.7 {1}	0.54 {1}
BahiaSul	0.3 {1}	5.43 {1}	0.33 {1}	0.11 {1}
Jari	3.4 {1}	35 {2}	1.3 {1}	< 0.5(short fibres) {1} < 1.0 (long fibres) {2}

Source: Firms' publications and complementary information 1996.
Ratings in {}: {1} = good; {2} = average; {3} = poor.
Abbreviations: ADT, air dried tonnes; AOX, adsorbable halogen organics (or chlorinated organics); BOD, biological oxygen demand; COD, chemical oxygen demand; TSS, total suspended solids.

TCF pulping processes, resulting in new or modified environment-related products: ECF and TCF pulp. Such innovative processes have also primarily been driven by the 'chlorine issue' debate and have represented an unprecedented technological change in the supposedly mature pulp industry. Brazilian firms have been quick to participate and innovate in the pioneering group of firms in the world pulp industry in these processes. This is illustrated in Table 7.4 that shows Aracruz Celulose SA among the pioneer firms in the world in developing and patenting the TCF production process. In doing so, firms have relied mainly on their internal technological capabilities and creativity, as well as their internal financial resources.

Their already strong capabilities in forestry biotechnology have been extended into environment-related targets. Such capabilities in biotechnology have represented the main basis for 'green innovation dynamics', opening up possibilities of further exploitation through process and product innovation currently under way in those firms. For such purposes, strategic partnerships with their traditional technological partners in Brazil (forestry research institutes and EMBRAPA – Empresa Brasileira de Pesquisa Agropecuária) have been intensified, and additional new partnerships with researchers at the leading edge of biotechnology developments in OECD countries have been established.

The main environment-related managerial and marketing innovative responses, involved increasing the importance of environmental departments in those firms' organisational structures, changing the focus of the environmental activities themselves from being local in scope and regulation-related to being global in scope and market oriented. Customers are clearly acting as new targets for environment-related information on products and processes in the pulp marketing activities. The product differentiation process introduced by the development of the ECF and TCF pulps, coupled with 'green marketing' initiatives, have represented a visible change in the dynamics of competition in the pulp markets, especially in Europe.

Table 7.4 Mills (pulp and paper) producing TCF pulp in 1992, worldwide

Species	*Country*	*Company*
Softwood	Sweden	Aspa
		Assi, Karlsborg
		NCB, Vallvik
		SCA, Ostrand
		Sodra Cell, Monsteras
		Sodra Cell, Morrum
		Sodra Cell, Varo
		Stora, Norrsundet
	Finland	Metsa-Botnia, Kaskinen
		Metsa-Botnia, Kemi
		Sunila
	Norway	Tofte
	Canada	Howe Sound
		Weldwood
Hardwood	Sweden	SCA, Ostrand
		Sodra Cell, Monsteras
		Sodra Cell, Morrum
	Finland	Metsa-Botnia, Kemi
		Metsa-Botnia, Kaskinen
	Spain	Ence
	Brazil	Aracruz
		Suzano (Paper mill)

Source: Moldenius (1992).

The pursuit of a 'green reputation' has also been observed as a growing concern amongst pulp firms. In order to succeed in targeting their customers for those purposes, eco-labelling and environmental management systems (EMSs) have, in recent years, received particular attention for their suitability in fulfilling such requirements.

The intensification of lobbying activities around the development of the EU eco-labelling scheme was observed among ABECEL firms. This represented an unusually reactive attitude by those firms when compared to their historical interactions with local regulations regarding industrial emissions. Such a reaction has been prompted by fears that the EU eco-labelling schemes might be used as technical barriers to exports from non-European pulp producers in European pulp markets.

Thus, at the same time as the Brazilian pulp export firms were carrying on pioneering investments towards cleaner production to comply with more stringent environmental standards and requirements, they were also lobbying to avoid the establishment of technical criteria that could place non-European producers at a disadvantage compared to their European competitors in reaching the pulp markets in Europe.

The changing nature of environmental pressures and firm responses

The analysis of those firms' recent environmental experiences demonstrates a clear shift in the nature of the environmental pressures in the world pulp industry, from strictly regulation-related pressures to new market-related ones. As a consequence, an important shift in the nature of the firms' environmental responses was also revealed, with the incorporation of green innovation into their economic and technological dynamics. Thus, strictly regulatory related pressures on this industry have historically led to passive technology absorption responses, aimed at compliance in a static way. In contrast, market-related pressures have led to dynamic technological and managerial innovation, clearly influencing their competitive dynamics in such an industry. Success and failure in such markets have been correlated with firms' capabilities to innovate, green innovation included.

Such a change in purchasing behaviour in the pulp and paper industry began in Europe from the end of the 1980s through pressures exerted by environmental activist groups (especially Greenpeace). From being geographically concentrated in Germany, such direct influences on the marketplace spread to various countries within and outside Europe through designing and implementing eco-labelling schemes, above all the EU eco-labelling schemes.[5]

During the long period of defining EU eco-labelling schemes for paper products, preventive approaches have been adopted throughout the leading firms in the pulp industry in countries supplying pulp to Europe. Coupled with such initiatives were investments in 'green innovation' carried out by those firms primarily in process changes (TCF, ECF and TEF pulping processes) and at the same time a renewed concern about improving marketing.

The EU eco-labelling process itself proved to take much longer than expected. After five years, criteria were actually defined and labelling schemes had become operational for only a few products. During the first half of the 1990s, the major influence exerted by EU eco-labelling on the leading pulp firms' activities became quite clear, not only as regards joint action for lobbying purposes on this issue but also in relation to the intensification of R&D aiming at anticipatory activity towards cleaner production. The final targets of such actions have been twofold: to enable firms to meet progressively more stringent emission standards underlying the eco-labelling schemes; and to satisfy new customer requirements for greener products, also linked to eco-labelling procedures.

Ironically, however, the great influence of EU eco-labelling schemes in stimulating investments in green innovation in firms in the pulp industry within and outside Europe in the early 1990s may well fail to continue to stimulate 'cleaner production' in the pulp industry in the near future. Such an observation leads to two important lessons with clear policy implications. First, the analysis carried out in the thesis pointed to the significant potential of the eco-labelling schemes in fostering green innovation as a step in the right direction of clearly affecting the demand side of the pulp industry. Second, the strong prospects for its failure[6] warrant further reflections on the regulatory approach to designing and implementing voluntary schemes like eco-labelling, as compared with those historically used to set emission standards. Specifically, further research and analytical effort should be directed to how to boost co-operation between firms and regulators aiming at cleaner production practices.

The environmental experiences of the Brazilian pulp export firms showed that Brazilian public policies played an important role up until the 1980s, particularly the financial policies of the Banco Nacional de Desenvolvimento Econômico e Social (BNDES), an organisation that has been a strategic partner of ABECEL firms from the outset. Indirectly, public policies for economic and technological development of forestry activities in Brazil also played a part. However, public policies for the environment, implemented from the 1970s onwards, played little role in fostering innovation, including green innovation. The Brazilian national environment system, however, has set the general guidelines for the environmental impacts of operations by firms. Environmental regulation in Brazil is ultimately defined at the local level, and is correspondingly firm specific.

The Brazilian national environmental system could be considered as up-to-date when compared to those operational in most OECD countries until the late 1980s. However, the developments in environmental regulation in Europe in the 1990s have created a relative gap in terms of the participation and perception of the leading edge of the current environmental debate between regulators in Brazil and their counterparts in OECD countries, especially in Europe. Even the more general information on the environmental debate in Europe had not spread among most regulatory bodies in Brazil.

There has been a recent effort to discuss and design green labels in Brazil; however, this has resulted directly from the initiative of firms in providing information and leadership to overcome national institutional fragility in the field. The most serious weaknesses have been the lack of financial resources and lack of emphasis on personnel training. The environment has not claimed the attention of long-run policy design to the extent observed for industrial and trade policies, which have fallen under the shadow of the emphasis on stabilisation policies. Despite the great effort put into the preparation of hosting the UNCED 'Earth Summit' in 1992, the environment has not figured as a priority for policy purposes in Brazil. These deficiencies have had an adverse impact on firms' institutional subsystems for coping with the increased environmental challenges posed internationally for Brazilian industries, especially the pulp export industry.

Despite such weaknesses in the Brazilian institutional system, and notwithstanding the fact that a large number of firms in OECD countries following different paths faced closure, Brazilian pulp export firms have, up to the present, proved themselves to be able to survive the difficulties represented by the renewal of environmental pressures in the world pulp industry, especially because they have been able to undertake 'green innovation'. Such an argument was supported by information provided in the research that revealed:

- Aracruz Celulose to be among the pioneering firms in the world pulp industry in developing TCF pulp processes; the fast generation and diffusion of ECF pulp processes in all ABECEL firms.
- The pioneering initiative of BahiaSul in being accredited under international EMSs (BS 7750 and ISO 14000), initiating a trend that will probably soon be general throughout the pulp and paper industry in Europe.
- Finally, and preceding the influences of eco-labelling on the industry, the pioneering innovative approach developed by Riocell to recycle solid waste. The latter firm's 'solid residue-free mill project' has a good opportunity to spread, at least in part, among the leading firms in the pulp industry, once landfill constraints come to receive increasing attention from regulators and local communities in different countries.

Innovation dynamics in the Brazilian pulp industry

After this analysis a final question can be posed in this chapter, as follows: what can be learned from such experiences to foster green innovation within firms? Many of the answers are implicit in what has been covered above. The detailed analysis of the internal innovation dynamics within those Brazilian firms revealed that technological capabilities accumulated during decades of continuous and purposive investments in research activities, especially in forestry research, were a primary strength. Until recently this pattern has not suffered any significant interruptions despite numerous changes in the board of directors, if for instance the experience of Aracruz Celulose is taken as an example. Personnel training and special attention to and support for international contacts involving research and technical personnel were seen to be a continuous and long-term policy in those firms, who were very keen to benchmark world-best practices in their industry and to anticipate trends considered as being strategic.

The analysis also reflected a progressive independence of those firms from the Brazilian state, as compared with the situation at their outset, as illustrated by their continuing investments in research despite the ups and downs of public policies for science and technological activities and their recent developments of capabilities in finance by operating in international stock markets. The traditional technological capabilities in forestry biotechnology were also shown to be a basis for internal innovation generation up to the present, for upgrading forestry research from its historical focus on wood yield to qualitative issues related to mitigating environmental impacts of the industrial processes, and for exploiting new market niches through the development of different paper grades. Biotechnology now underlies the whole pulp production process, and research was observed to be still under way, which seems likely to lead to further technical change in the industry. Continuous improvement in emissions will probably be observed in those firms in years to come. Special efforts are still needed to reduce air emissions of sulphur, even though at present the air emissions of those Brazilian firms have been about average compared to the world pulp industry.

Some comments should also be made about the observed firms' perceptions concerning environmental regulation and the existence of emission standards. The historical environmental experiences of the Brazilian pulp export firms showed no strong reactions against the stringency of local regulation. According to these firms, their general attitude, despite differences in style, is that the law is to be obeyed and they do so. Emission standards in many cases can be used as safeguards against casual complaints, because they can provide technical support that their activities are being carried out in accordance with established standards. Problems tend to occur when there are changes in the law, when the main alleged problems that emerge include: legislation is designed without technical expertise, because of lobby pressures, creating technical distortions in production, so a more scientific basis is required; or because their participation in the discussion fora are denied despite their activities being highly influenced by such legislation. Those problems, for instance, were alleged as being present in the design of EU eco-labelling schemes and were the main reasons underlying the well known criticisms by Brazilian pulp firms of such schemes.

The continuous pattern of investments in R&D and in expansion of production capacity have been important roots for the pattern of continuous growth and development observed in this industrial branch in Brazil, despite the cyclical variations in the national economy and the discontinuities of long-term public policies. While the early development of those firms relied heavily on support from public policies, in the past decade firms have had to rely on their internal financial and technological resources in facing the renewed environment-related challenges.

Building upon the strong technological capabilities in forestry research present in those firms' institutional subsystems (in forestry research institutes, EMBRAPA and the firms themselves), the partnerships with foreigners provided invaluable initial technological learning about large-scale pulp production, finance and access to the international pulp market.

However, additional focus and capability development is required in order to maintain the financial and technical capability of firms to innovate in general, including green innovation. These relate to financial issues on the one hand and to the maintenance and reinforcement of research capabilities and marketing on the other. In 1992, alongside the increase in environmental pressures, the world pulp industry experienced its deepest price downturn in history, imposing additional constraints on financing firms through eroding profitability. After a recovery from the end of 1994 up until the beginning of 1996, prices of market pulp have experienced another downturn, revealing that such cycles in the industry have become shorter and deeper. Bearing in mind that a similar price pattern has not been so evident with regard to final paper-product prices, great effort needs to be directed by the industry towards better understanding of the emerging dynamics of the pulp trade, in order to design long-term corporate strategy more effectively. This means the need for further development of capabilities in trade and pulp marketing, with possible direct effects on firm profitability, generating internal financial sources for further investments.

Linked to the price crises and organisational restructuring in leading firms in the pulp industry in Scandinavia, another important issue affecting the technological capabilities of Brazilian pulp firms relates to policies on cost cutting. The research revealed that most of the firms have been experiencing significant downsizing. Re-engineering programmes were under way in two of them, and subcontracting programmes have been used extensively. The analysis revealed that renewed concerns about total costs are legitimate. Sweden and Finland, competitors in this market, did the same a few years ago. The important issue to be raised in this regard is the need for care in such cost reduction initiatives, in order to avoid the loss of those technological capabilities that have accounted for the successful economic trajectory of the Brazilian market pulp industry up to the present day. The main challenge in this respect is to reconcile reasonable costs with the maintenance and reinforcement of firms' technological capabilities in forestry and biotechnology, and with extending personnel training to building capabilities to operate in the international stock market (already initiated in some of these firms), and in trade to increase direct sales and customer technical services to face the trend towards increased concentration in the international pulp markets.

With regard to the Brazilian system as a whole, the research revealed a need for the government to be able to overcome the immediate necessity of stabilisation policies, in order to be able to renew attention and investment in long-run integrated policies in trade, science and technology, and the updating of the national system of the environment.

Conclusion

The analysis has led to four main conclusions, discussed in turn as follows. The thesis did not advocate the superiority of market-based mechanisms over regulation or vice versa. In fact it was found that both can lead to cleaner production. However, as a first main conclusion, the analysis clearly showed that historically the effect of regulation upon those firms has led to regulation compliance in a quite passive way, by purchasing equipment and services from the shelves of the pollution-abatement supply industry. As soon as in the past decade environmental issues intensively affected their markets by changing customer purchases towards greener products in Europe, those firms internalised environmental issues into their core technological and marketing activities and these began to play the major role in such a green innovation dynamics, as illustrated by the development of ECF and TCF pulp and the changing focus of biotechnology research within those firms. According to the results of this research, firms make green innovation endogenous to their core technological competencies and marketing activities when it becomes an important requirement for maintaining and reinforcing market positions despite the general recognition of the political pressures underlying this phenomenon.

Thus, it is argued that the main means for enhancing green innovation in industrial firms, according to the recent pulp industry experience just analysed, is by stimulating green consumerism on the demand side, and acting to strengthen firms' technological partnership possibilities by tackling weaknesses in their 'sectoral subsystems of environmental innovation' (SSEI)[7] on the supply side. A better understanding of 'green innovation dynamics' within firms is crucial to seeing how innovation can promote cleaner production processes in industry. In making this argument, the chapter also argues that 'linear' approaches to the regulation–innovation interface, which by and large have underpinned current theoretical approaches in both the economics and the business field, have failed to establish the dynamics of environment-related innovation in firms. The main reason is that they have emphasised factors external to the firms (which envisage the driving forces as inputs and firm responses as outputs), while processes internal to the firm of generating environmental responses have remained under-investigated.

To fill this gap, the use of the proposed analytical tool of SSEI[8] for systemic analysis of the 'green innovation dynamics' in firms might represent an insightful option. Building on theoretical contributions provided by the 'national systems of innovation' approaches of Freeman (1988), Lundvall (1992) and Nelson (1993) and on the 'systems of production' approach of von Tunzelmann (1995), the SSEI represents a theoretical attempt to integrate the environment into the technological debate on industry. It stresses the need for increasing the analytical focus on the firm level, to allow a better understanding of 'green innovative dynamics'.

The SSEI places the firm at the epicentre of the analysis of green innovation dynamics, as illustrated in Fig. 7.6. Its geographical delimitation will depend on whether the firms' economic activities are geographically local, regional or global. By identifying the sources and nature of environmental challenges *vis-à-vis* firms' internal production capabilities (technological, financial and marketing capabilities) to cope with them, such a framework can provide a basis for mapping firms' needs with regard to partnership establishment for such purposes. As a consequence, such analysis also provides a basis for defining environment-related corporate strategies. At the macro-level, the mapping of actual and required partnership establishment by firms

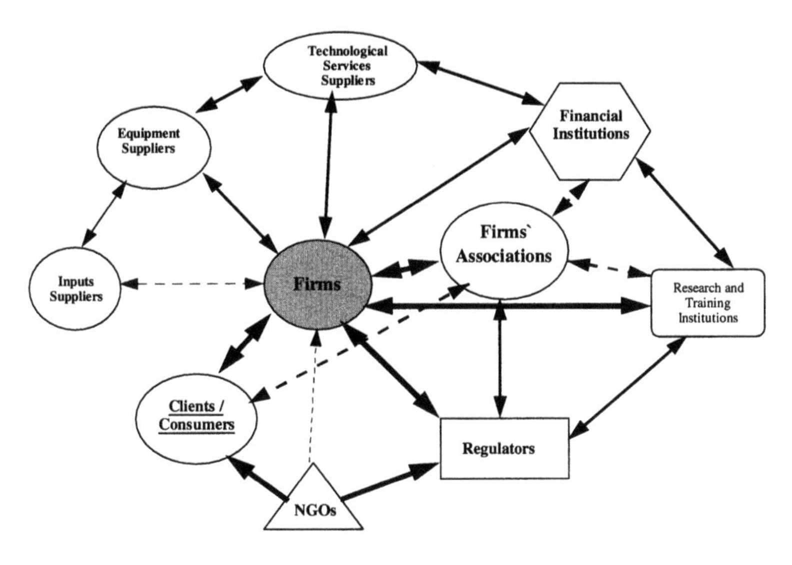

Figure 7.6 Sectoral subsystem of environmental innovation.

for environment-related research may reveal weaknesses and strengths in the firm.

At the firm level the SSEI can contribute to a systemic analysis of environmental challenges it faces and guide the definition of environment-related corporate strategies. At the macro-level the SSEI can contribute to the identification of weaknesses and strengths in the sub-systems under analysis, providing a basis for design and implementation of public policies at national or sub-national levels, aiming both at tackling such weaknesses and fostering cleaner production practices in the industry.

The SSEI approach thus departs from the linear regulation–innovation models in giving maximum emphasis to the proactive role of customers, and through such feedback identifies the dynamic environment-related interactions. This helps to detect the strengths and weaknesses in such sub-systems so as to provide a basis for action for improving both corporate environmental strategies and the design and implementation of public regulation. It fits in non-linear regulation–innovation models, as illustrated by Fig. 7.7.

As a second main conclusion it is stressed that the introduction of market mechanisms in European environmental regulation in the past decade has promoted a change in the nature of environmental pressures and has fostered innovation within Brazilian firms in the pulp industry.

European markets have represented a main destination for the ABECEL firms' pulp, with the exception of Cenibra. This last firm, due to a shareholder agreement, has never had less than 50 per cent of its production marketed in Japan by its Japanese shareholders. In Europe, Cenibra has

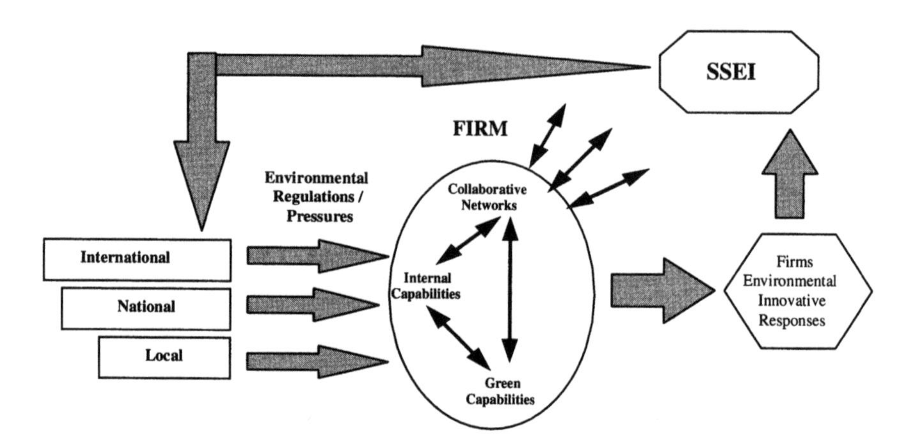

Figure 7.7 Dynamic regulation/innovation model.

marketed about 19 per cent of its exports. Jari, Aracruz, Riocel and BahiaSul, however, have marketed 54, 35, 30.5 and 29 per cent, respectively, in Europe, as shown in Table 7.5. All those firms emphasised the predominance of clients' demands for ECF and TCF pulp from European clients.

Although European markets have provided most of the environmental challenges, it has not been Jari that has invested more in environment-related innovation, despite being proportionately the largest exporter to Europe. This corroborates the conclusion that, despite market signals for firms to undertake environmental innovation, markets alone do not provide the means for so doing. Environmental innovation is thus dependent on firms' internal technological, financial and marketing capabilities, combined with firms' partnership establishment possibilities within their SSEI. It was in Aracruz, where the combination of market pressures and strong internal technological and financial capabilities were observed, that the

Table 7.5 Brazilian bleached kraft pulp destinations, 1993 (tonnes)

Firm	*Own consumption*	*Domestic market*	*Export*	*Total*	*European markets*
BahiaSul	88,588	50,031	306,412	445,031	29[a]
Aracruz	0	88,027	931,571	1,019,598	35[a]
Cenibra	0	36,427	339,534	375,961	19[a]
Jari	0	24,405	111,525	135,930	54[a]
Riocell	0	76,232	160,316	236,548	30.5[a]
Others	1,105,691	202,219	6,735	1,314,645	n.d.

Source: ANFPC (1994); Author's calculations.

Note

a Data (%) for exports in 1994. Fieldwork interviews, 1995.

intensity of investments and positive results with regard to cleaner production practices has been most intensively taking place in the past decade.

As a third main conclusion the research revealed that firms' environmental responses in the pulp industry have changed in nature in the past decade – from being 'end-of-pipe', local and disconnected from firms' technological strategies, to being global, synchronised and related to their core technological and marketing activities.

Brazilian pulp export firms' historical environmental experiences were somewhat firm specific, driven by local issues, directed towards regulatory compliance and local communities' acceptance. Each such experience very much depended on the timing of the investment, the geographical location, the sociological characteristics of the local community and the economic structure of each locality. With few exceptions each experience and environmental response provided did not generalise their practices to counterpart firms. The main means of responding to environmental pressures was by purchasing equipment and services from the pollution-abatement supply industry. In the past decade, however, in Brazil and in the main OECD pulp and paper producers, a synchronised process of investments and research efforts towards product and process innovation and cleaner production practices in general has been observed.

Finally, the research also concluded that what enables a firm to perform as an innovator, as distinct from being a mere technological absorber, is its internal research capabilities. Such a conclusion was especially corroborated by the experience of Aracruz Celulose in developing TCF pulp, together with a few other pioneering firms in terms of green innovation in the international pulp industry.

Last but not least, it is important to stress that, by putting the focus on the firm level, this chapter has not intended to disregard the role of government in tackling environmental challenges or the important role played by environmental pressure groups. On the contrary, governments have crucial roles to play in facing the sustainable development challenge. However, industrial green innovation within firms can also play a very positive and important role in combining wealth generation with improving the quality of life. Thus, an understanding of firm sensitivity to environmental issues and the main factors underlying capabilities and decisions to invest in cleaner production is extremely important.

Notes

1 This chapter draws on the main findings of the author's PhD thesis ('Dynamic Capabilities for Cleaner Production Innovation: the Case of the Pulp Export Industry in Brazil') presented at the University of Sussex in September 1997.

2 For detailed data on the pulp and paper industry in Brazil see ABECEL and ANFPC (various issues).
3 Standard pulp is bleached by using chlorine.
4 For a summary on the green and competitiveness debate, see Howes *et al.* (1997); Porter and van der Linde (1995) and Walley and Whitehead (1994).
5 For more detail on eco-labelling schemes, see Bristow (1994).
6 Such failure is based on the low level of formal applications by paper firms and the formal decision of the Confederation of the European Paper Industries (CEPI) to withdraw its support from such schemes, changing its support to environmental management systems instead. (Interview with Ms Annick Carpentier, Manager for Environment, at CEPI, Brussels, 17 December 1996.)
7 SSEI is an analytical tool proposed in that research, that, by placing the firm in the epicentre of a diagram, intends to trace its main economic and institutional relationships for production and environmentally related innovation processes.
8 For details on SSEI, see Dalcomuni (1997).

Bibliography

ABECEL (1993) *Selo ecológico da CEE: Posição e ações do setor de celulose e papel.* Rio de Janeiro (mimeo).

ABECEL (various issues) *Annual Report.* Rio de Janeiro.

ANFPC (various issues) *Revista Celulose e Papel.* São Paulo.

Aracruz Celulose SA (various issues) *Annual Report.* Rio de Janeiro.

BahiaSul SA (1995) *ISO 9002/BS 7750: BahiaSul Recebe Certificado de Qualidade e de Meio Ambiente.* Mucuri (mimeo).

BahiaSul SA (various issues) *Annual Report.* São Paulo.

BNDES (1993a) *Papel e Celulose de Mercado – Panorama 1980/1992.* Rio de Janeiro.

BNDES (1993b) *Informe sobre estratégias setoriais e empresariais: Meio Ambiente, Regulamentação e estratégias empresariais.* Rio de Janeiro.

BNDES (1996) *BNDES Sectorial 3.* Rio de Janeiro.

Bristow, P. (1994) 'The European Community Eco-Labelling Scheme', in *Life Cycle Management and Trade.* Paris: OECD.

Business Council for Sustainable Development (1992) *Nosso Projeto Comum – Uma perspectiva empresarial para o Desenvolvimento Sustentável na América Latina.* Genebra.

Cenibra SA (various issues) *Annual Reports.*

Collins, L. (1994) 'Environmental Performance and Technological Innovation: the Pulp and Paper Industry as a Case in Point', *Technology in Society*, 16 (4): 427–46.

Dalcomuni S.M. (1990) *A implantacao da Aracruz celulose no Espirito Santo: principais interesses em jogo.* Rio de Janeiro: CPDA/UFRRJ, Dissertacao de mestrado.

Dalcomuni S. M. (1997) *Dynamic Capabilities for Cleaner Production Innovation: the Case of the Market Pulp Export Industry in Brazil.* Unpublished DPhil thesis, SPRU, Brighton: University of Sussex.

FAO (1995) *Pulp and Paper: Perspectives Towards 2010 and Beyond.* Rome: Food and Agriculture Organisation.

Freeman, C. (1988) 'Japan a national system of innovation?', in G. Dosi *et al.* (eds), *Technical Change and Economic Theory.* London: Pinter.

Howes, R., J. Skea and B. Whelan (1997) *Clean and Competitive? Motivating Environmental Performance in Industry*. London: Earthscan.

IIED (1996) *Towards a Sustainable Paper Cycle*. London: International Institute for Environment and Development.

Jaako Pöyri (1996) *Assessment of Bleaching Technology and Emission Control in the Pulp Industry*. Sub study series 8. *Project Towards a Sustainable Paper Cycle*. London (mimeo).

Jari Celulose SA (various issues). *Annual Report*.

Lundvall, B.A. (ed.) (1992) *National Systems of Innovation: Towards a Theory of Innovation and Interactive Learning*. London and New York: Pinter.

Moldenius, S. (1992) 'The status and future of totally chlorine free bleached kraft pulp'. Brussels: paper presented at the PPI Market Pulp Symposium 8.

Nelson, S. (ed.) (1993) *National Innovation Systems: a Comparative Analysis*. New York: Oxford University Press.

Porter, M.E. and C. van der Linde (1995) 'Green and Competitive: Ending the Stalemate', *Harvard Business Review*, Sept.–Oct.: 120–34.

PPI (1996) *International Fact and Price Book*. Brussels: Pulp and Paper International.

Price, D. (ed.) (1994) *The Impact of Environmental Legislation on the European Paper Industry: an environmental legislation handbook*. Tonbridge: Benn Publication Ltd.

Riocell SA (various issues) *Annual Report*.

Tunzelmann, G.N. von (1995) *Technology and Industrial Progress: the Foundations of Economic Growth*. Aldershot: Edward Elgar.

UNIDO (1993) *Industry and Development Global Report*. Vienna: United Nations.

Walley, N. and B. Whitehead (1994) 'It's Not Easy Being Green', *Harvard Business Review*, May–June: 46–51.

8

ENVIRONMENTAL PERFORMANCE AND TRADE LIBERALISATION IN THE MEXICAN TEXTILE INDUSTRY[1]

Flor Brown

Introduction

Recent literature (OECD, 1995; Sorsa, 1994; Low, 1992) has pointed out the impact of increasing globalisation on the levels of industrial pollution. In this research we hope to examine to what extent industrial development, led by industrial integration into the world economy, has contributed to or reduced environmental damage in the case of the Mexican textile industry. Although this industry is not characterised by as high a degree of globalisation as other industries in Mexico, the wide differences among firms in terms of export shares may throw light on the determinants of environmental behaviour.

The results of this investigation confirm that many firms have difficulties in undertaking investments to confront the adverse effects that textile production processes have on the environment and suggest greater heterogeneity in terms of the firms' environmental performance than in other industries in Mexico, for example, the chemical industry. The analysis of 19 firms shows that there are statistically significant differences in the environmental performance between firms (a) with high and low levels of exports; (b) with recent and old technology; and (c) large and small firms.

There are relatively few firms that have a proactive environmental response, and progress towards more complex strategies at the firm level as a whole is still incipient and found in only a few firms. However, a positive association between such proactive behaviour and export orientation, level of technology and large firms was observed. Thus, the hypothesis that the requirements of the international market force export firms continuously to adopt cleaner technologies, widen their production scales and reduce inefficiencies in production is valid in the case of a few firms in the textile industry.

This chapter is divided into three sections. In the first section we describe aspects that characterise the industry: recent economic performance, the external sector, the production process and the impact on the environment. The second section analyses the responses that 19 textile entrepreneurs gave in a survey undertaken in the metropolitan area of Mexico City and in the state of Puebla. In the final section, hypotheses related to the trade opening and environmental performance are verified, and conclusions arising from the research are presented.

Industrial characteristics

Cloth production in the Mexican textile industry is composed of three segments: yarns and woven goods of soft fibres, yarns and woven goods of hard fibres, and other textile industries. Together, the production of cloth accounts for 7 per cent of GDP and 5 per cent of employment in manufacturing, while contributing 55 per cent of GDP and 50 per cent of employment in the textile sector as a whole (excluding the footwear industry).

Recent aggregate data for the textile industry point to signs of decline after trade was liberalised. This is due to a number of factors, most importantly the fall in investment in the sector and the increase in imports. The downturn in production can be seen in Table 8.1; between 1988 and 1993 production decreased at an average annual rate of 5 per cent. After a long period of crisis between 1994 and 1997, production started to increase at an average annual rate of 8 per cent.

The textile industry's slump is not only reflected by the drop in production, employment also fell throughout the period (Table 8.2). Employment fell at

Table 8.1 GDP of Mexican cloth production

	Constant pesos of 1980 (millions)
1988	4,885
1989	4,938
1990	4,832
1991	4,503
1992	4,021
1993	3,731
1994	3,812
1995	3,772
1996	4,363
1997	4,741

Source: INEGI, *Cuentas Nacionales*.

Table 8.2 Employment in the textile industry

	Number
1980	175,344
1981	179,496
1982	170,714
1983	166,723
1984	167,460
1985	171,852
1986	168,178
1987	169,092
1988	168,428
1989	169,999
1990	164,002
1991	154,540
1992	139,738
1993	122,430
1994	108,039

Source: INEGI, *Cuentas Nacionales*.

an average annual rate of 0.4 per cent between 1980 and 1985, and 5.4 per cent between 1986 and 1994.

Owing to the low degree of utilisation of available production capacity until the mid-1980s, gross investment in physical capital assets shrank. Investment throughout the period never regained its 1980 level. From 1980 to 1985, investment in physical assets fell at a rate of 21.4 per cent and 12.7 per cent between 1986 and 1994. Most firms in the sector have not undertaken significant investment in machinery; the downward tendency has persisted over 15 years (1980–94), with drops of 12.7 per cent between 1980 and 1985, and 9.0 per cent between 1986 and 1994.[2] These trends in investment show that, for a relatively important group of textile firms, there is a technological lag in machinery used in production processes. Padilla (1994) points out that the most important technological lag exists in the dyeing and finishing of cloth, due to the use of obsolete machinery, resulting in high costs and technical inefficiency in the use of chemical and energy inputs. Such technological lags are an important factor contributing to the negative impact which the industry has on the environment.

The investment lag in machinery is explained in part by the difficulties firms face in obtaining finance. To have access to credit, it is now necessary to have both adequate size as collateral, and for the firm to show export capabilities or to negotiate joint ventures with foreign firms (Padilla, 1994).

Local firms have been unable to meet internal consumer demand, and thus trade liberalisation has brought greater imports of textile products. As Table 8.3 shows, imports have grown rapidly at 24 per cent per annum between 1988 and 1997.

Table 8.3 Imports and exports of the textile branches

	Imports (US$ million)	*Exports (US$ million)*
1980	108.6	407.7
1981	128.3	410.1
1982	88.6	205.1
1983	24.9	255.2
1984	50.7	401.4
1985	76.2	229.0
1986	77.7	330.5
1987	105.2	441.1
1988	253.7	461.4
1989	391.0	432.3
1990	507.5	378.2
1991	701.4	445.7
1992	893.8	448.5
1993	1,029.4	559.4
1994	1,437.2	674.7
1995	900.3	1,233.6
1996	1,223.8	1,323.8
1997	1,729.1	1,571.7

Source: *Sistema de Cuentas Nacionales de México.*

Although some firms have been able to increase their exports in the period following trade liberalisation (Table 8.3), the average annual rate of growth in exports of textile products between 1988 and 1997 (15 per cent) was lower than the rate of growth of imports. These data show that, in spite of the efforts of entrepreneurs to adjust their production capacity to foreign demand, competition in the international market prevented exports from growing at a faster rate.

Summarising, the statistics from this segment of the textile industry show drops in levels of production since the trade opening. Imports of these textile products have risen considerably and investment, principally in machinery, has dropped. Within the past few years, no important investments in technology or in efforts to expand markets have been recorded. These aggregate data suggest that many firms that produce cloth may possibly be having great difficulty undertaking investments that would permit them to confront the adverse effects that the textile production processes have on the environment.

Environmental impact and the production process

The textile industry consumes large quantities of water in its production processes and because of the chemicals used in the process, wastewater is difficult to treat. The industry accounts for over 3 per cent of all industrial wastewater discharges in Mexico, making it the seventh most important industry in terms of water pollution (CNA, 1994; Table III.5). The production processes for cloth start from fibres that may be natural (wool, cotton) or synthetic. Interviews were carried out in firms that use fibres such as cotton as well as synthetic fibres.

The process begins with the preparation of fibres. Fibres are mixed, depending on the requirements of the cloth. These operations are carried out in opening machines and fulling mills. The opening machinery fluffs out and separates fibres in tufts that are so small that the heavier impurities in the fibres can be removed. At the same time distinct fibres are being combined. The fibres are then carded and combed, and pulled; the last step in this part of the process is torsion. Up to this stage the pollution problem was one of lint and dust for which firms have installed filters to reduce air pollution.

Before the skeins can be used for the weaving of cloth, substances such as starch are added to make the fibres stronger. This is a three-stage process: pasting, drying and rolling. These stages generate 30 per cent of the pollution of wastewater, given that acids and enzymes are used, in addition to caustic soda, chlorine, detergents, etc.

Once the cloth comes off the looms, it goes through a number of operations: elimination of threads and weaving defects; modification to the fabric's appearance, e.g. treatment to improve the appearance of the fabric (gloss or frieze, etc.); chemical dressing through impregnation with liquids composed of starch, dextrin and other substances that lend body to the cloth; special finishing to make the fabric wrinkle-free, deformation-free, impermeable, etc.

The last stage is the dyeing and finishing of the cloth. Once the products have been dyed, they are washed with water and left to dry in the open air or in dryers. Sometimes colours are made brighter and special patterns are obtained by submitting the dyed products to an oxidising atmosphere, with steam jets, very hot soapy washes, etc. Lastly, the cloth is given the desired texture in terms of softness, brilliance and appearance. A great variety of chemicals is also used such as starches, dextrin, chlorine-based substances, softening agents, etc. This phase accounts for between 20 and 40 per cent of the total water pollution. The phases of the production process are illustrated in Fig. 8.1 (see Gurnham, 1965)

From the questionnaires we can surmise that the principal forms of pollution derived from the production process are the emissions into the air from the weaving process, and water pollution from the dyeing and finishing of cloth. Firms have concentrated their attention on these problems. Air pollution has been reduced through the use of filters, while most investments and efforts have centred on water pollution.

Only a few firms visited have invested in new, more efficient machinery to modify production processes in an effort to reduce pollution. Most firms

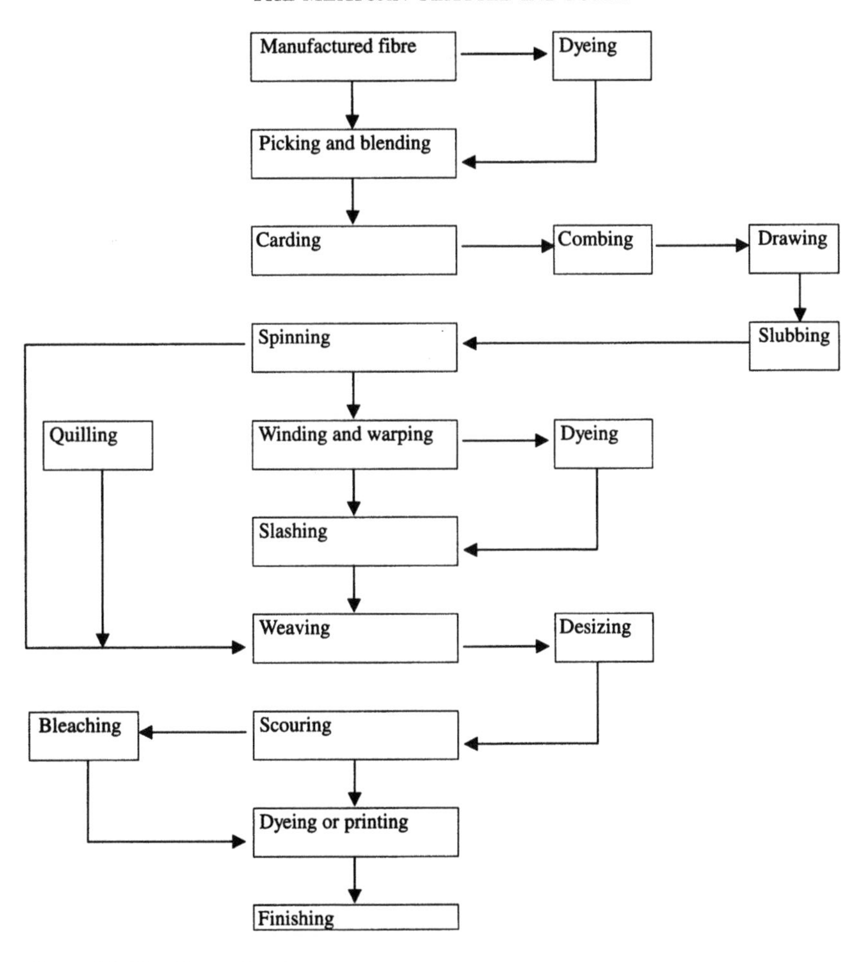

Figure 8.1 Textile production process.

simply reduce the use of chemicals and substitute colouring agents. Few firms have a water-treatment plant or recycle the water. Plant efforts have centred principally on the problem of wastewater treatment. Only nine plants recycle their solid and liquid residuals and have wastewater treatment facilities. The discharge of 58 per cent of the plants goes into the public drainage system, while five plants discharge directly into a river. Seven plants place their hazardous wastes in confinement centres.

The prevalent use of chemicals in dyeing both fibres and cloth, as well as in the finishing process, has led plants to modify inputs, this being the main effort made at improving the environment. Changes in equipment and in production processes for environmental reasons were less common and of a relatively minor nature. Two firms reported that they were introducing water treatment, and another two that they had stopped discharging wastewater into the river and had begun to recycle water.

Environmental performance

The aim of this section is to provide a picture of the environmental performance of the Mexican textile industry based on a survey of 19 firms carried out in 1998. The firms were visited and interviewed on the basis of an extensive questionnaire covering the general background of the firm (size, ownership, exports), environmental policies, environmental programmes and activities, environmental management, environmental drivers and environmental impact. The analysis of the questionnaires presented next shows a great heterogeneity between the companies of the sample in these aspects and in their overall environmental performance.

Characteristics of the sample

The sample was designed to include different-sized firms in order to analyse different patterns of environmental response. With the exception of one firm located in the state of Aguascalientes, the rest are located in the textile area of Mexico City and in the state of Puebla, where 25 per cent of the establishments and 50 per cent of all textile production are found. Firms were selected from the entrepreneurial directories of the Chamber of Textiles of Mexico City and the state of Puebla. From these directories a sample was chosen of 60 firms, of which 19 agreed to answer the questionnaire. In spite of the fact that the firms were randomly selected from the directories, the final sample is possibly biased (as will be seen later on) in favour of those firms with a particular interest in resolving environmental problems caused by their production processes. No information exists for those firms that did not permit us to visit their plants, and so it is difficult to evaluate their environmental compliance. It is likely that they refused to participate in the survey because they do not assign much importance to environmental issues and are either just complying with environmental norms or even possibly failing to do so.

Figure 8.2 shows the distribution of the different sizes of the firms visited. It was possible to cover different-sized firms with different characteristics. It is worth noting that none of the firms had, at the time of the visit, production of less than one million pesos. Two firms withheld sales information.

The share of foreign investment in the textile industry is not significant. According to the industrial census of 1994, in the textile sector, firms with foreign capital accounted for only 20 per cent of the gross value of production and thus only two firms with foreign capital were included.

The firms in the sample included three which belonged to a large industrial group with interests in a number of sectors, including chemicals, as well as

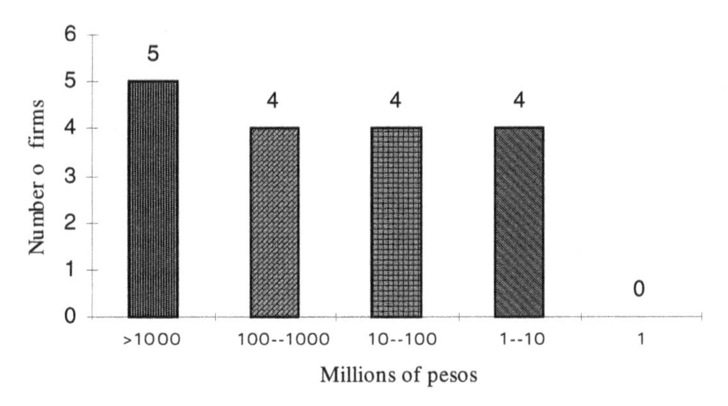

Figure 8.2 Firm sales. Source: Author's figures based on information provided by firms.

textiles. Two other companies were part of a major textile group. The remaining firms were independent manufacturers.

As mentioned, the textile industry's exports have increased in recent years, but there are relatively few firms with high levels of exports. In the sample, only three firms exported more than half of their production, while a further five exported between a quarter and a half. Five firms do not export at all and two more export very little (less than 5 per cent) while the rest export between 5 and 25 per cent of their production.

Firms' environmental policies

Compared to the chemical or steel industries in Mexico, few firms in the textile industry have an environmental policy. In our sample only five firms had a written environmental policy. The most important aspects that are included in the environment policy of these five firms are the commitments to reduce wastes and consumption of resources through recovery and recycling, continuous improvement in environmental performance and adherence to environmental laws and regulations.

The written documents on procedures that firms have available deal mainly with the effects of an environmental emergency in the plant. Ten firms have such documents and seven have them for the disposal of industrial wastes. There are few firms that have drawn up other types of documents regarding the effects of an emergency on the local community or the environmental impact of new projects. As shown later, firms with one plant have been able to solve, to a certain degree, the environmental problems of their plants, without having to draw up a written environmental policy.

Looking at environmental policy as a whole, it was possible to distinguish two groups of firms. The first group was made up of five firms which had a clear and explicit environmental policy. Of these, three plants were part of an important group of firms, and two plants belonged to a set of plants involved in cloth finishing. The remaining firms essentially lacked any environmental policy.

Programmes and activities

The most important problem faced by the textile industry is water pollution, so it is not surprising that the environmental programmes of the firms surveyed are especially concerned with reducing the discharge of liquid effluents, improving the efficiency of energy use and reducing solid waste. In twelve firms these programmes involve quantitative goals. In ten, goals are to be met in the short or mid-term, and in six the goals have been almost fully reached.

Other environmental programmes include training, environmental audits and investment. Training programmes in environmental matters are given to staff in charge of the environment and to workers. It is thus probable that workers have a limited participation in the culture and organisation of environmental matters. Training typically covers waste minimisation, treatment of water before it is discharged and the control of air pollution. In twelve firms voluntary environmental audits have been carried out, ten firms have undertaken environmental impact studies and eight carry out activities with the local community.

In order to obtain an overall picture of the extent of environmental programmes and activities undertaken by different firms, a quantitative exercise was carried out which assigned points to the various programmes undertaken with a maximum score of 7. Three groups of firms emerged. One group had three firms with scores greater than 5, which are members of the group of firms mentioned above; a second group of seven firms with scores between 3.1 and 5; and the last group of nine firms, with very few programmes, with scores between 0 and 3 points.

New investment in equipment and machinery by the firms surveyed has been fairly limited, especially in the years following the trade opening. This is in line with the experience of the textile industry as a whole as was pointed out earlier. This is most clearly indicated by the average age of machinery as calculated by the firms, which was 15 years. Not surprisingly, in view of the low levels of investment in the industry, investment aimed at improving the firms' environmental performance has been extremely limited, the most significant cases being two firms which introduced waste water treatment.

Environmental management

Interestingly, in terms of environmental management, thirteen firms claimed to have established a formal environmental management system, but only nine firms reported having an environmental management manual. Only five firms reported that they had established indicators of environmental performance, although most firms (73 per cent) claimed to collect environmental information systematically and the remainder collected information in an *ad hoc* manner when needed.

Only three firms grant incentives to improve the environmental performance of their employees. Most firms stated that fulfilment of environmental performance requirements is an employee's duty and therefore it is not necessary to grant incentives in this regard. Firms find it of little interest to publish information about their management of environmental matters. Only five publish this information. Firms are more concerned with establishing procedures to assure that all employees are aware of environmental effects. Nine firms have established these procedures.

In terms of environmental management, an interesting aspect is the use of specific indicators. Not all plants have a record of their use of fuel, electricity and water per unit of production. Only nine have indicators per unit of production for the use of electricity and water, twelve have them for fuel and eleven take readings of residual water. As for the monitoring of air and water, fourteen plants take samples at regular intervals and have statistical series on BOD, COD and suspended solids; the remaining plants take samples only when required, and do not have previous statistics on hand.

Firms were evaluated on the basis of the comprehensiveness of their environmental management systems and their monitoring of environmental performance. The maximum score of 20 was achieved by the three firms which belonged to the same large corporate group. Of the remaining firms, three more scored between 10 and 15, indicating a significant development of environmental management; nine scored between 5 and 10, showing incipient development; and four scored 5 or less, reflecting the fact that they had not yet put in place an environmental management system.

Environmental impact

So far attention has focused on the policies, programmes and management systems that have been applied by the firms covered in the study. However, it is also important to attempt to evaluate the actual environmental impact of the firms, both in terms of meeting Mexican norms and standards and in terms of achieving reductions in their impact on the environment over time. Given the characteristics of the textile production process for firms within the sample, the most important sources of environmental improvement are water treatment and secondly recycling.

In terms of their own evaluation of their environmental performance, seven firms claimed to exceed Mexican legal requirements. These firms have characteristics that should be mentioned. Three of them belong to an important group made up of chemical and textile firms. The corporate offices of these firms began to express concern regarding environmental matters in 1978. Since then they have linked their commitment to reduce pollution with the modernisation of the firms within the group. The group has accumulated knowledge and experience in environmental control, thus permitting firms to modify their environmental performance. Another two firms belong to an important group of firms that finish cloth. The quality in the finishing process used by these firms has been widely recognised in the US market. The firms have state-of-the-art technological installations and a modern plant for water recycling and treatment. As in the previous case, both the concerns about introducing new technologies and environmental matters have been in the forefront of the head offices' export strategy for these firms.

Seven firms from the rest of the sample claim to comply explicitly with national environmental norms, while the remaining five firms normally comply but have on occasion failed to meet the norms. However, a somewhat different picture emerges when firms were asked whether they had been in breach of environmental regulations in the past. Twelve firms reported that action had been taken against them (mainly fines or written warnings) including several of those who described themselves as exceeding national norms. Usually, sanctions are imposed for breaking water standards. The Federal Law of Rights in Matters of Water (1997) specifies the characteristics of water discharges. Fines can be imposed for two reasons: either for not covering charges imposed for water discharges or because the discharges are not within specified guidelines for pollutants established by the law. Penalties often require the payment of a fine or, in extreme cases, the firm can be sanctioned with the temporary closing of the plant. These sanctions imposed on the firms suggest that although most of them claim to meet environmental norms they have not always been within the norm.

In terms of improving performance over time, firms can reduce their environmental impact through more efficient use of resources and through reducing emissions. Ten plants undertook efforts to reduce the use of fuel, and eight consumed less electricity. However, these changes were probably mainly motivated by economic rather than environmental considerations. In spite of the fact that the industry uses vast amounts of water, only a few plants (six) reported a drop in the use of water.

In terms of emissions and discharges, again given the nature of textile production, water pollution is the most important indicator. Only a minority of firms (seven) reported that they had reduced the quantity of wastewater in recent years. Six firms reported that the quality of water discharges had improved in terms of BOD, COD and suspended solids. Seven firms were able to show reductions in air emissions, and five firms reduced the amount of solid waste generated.

Environmental drivers and obstacles

The drop in production within the textile industry since trade was liberalised has led firms to decide against improving their environmental performance on their own initiative. Therefore, by far the most important factor leading firms to take measures to improve their environmental performance has been governmental regulation. This is followed by pressure from local communities, which was an important additional factor for a few firms (Fig. 8.3).

Firms faced a number of obstacles that have prevented them from improving their environmental performance. Looking at obstacles that are external to the firm, the most commonly mentioned factor was the high cost of the equipment required. Other factors considered important by some firms were government policies, the lack of incentives, and competitive pressures (Fig. 8.4). Also mentioned was inadequate local infrastructure for the joint treatment of wastewater, which translates into high costs and inefficiencies for individual firms.

Firms were also constrained in achieving better environmental performance by a number of factors internal to the firm. A majority of firms cited the fact that they had other more urgent priorities as a major obstacle.[3] Another factor mentioned by a number of firms was the lack of adequate funds (Fig. 8.5).

The picture that emerges from the survey is that textile firms have adopted environmental measures predominantly in response to external pressures, particularly as a result of government regulation and in certain cases

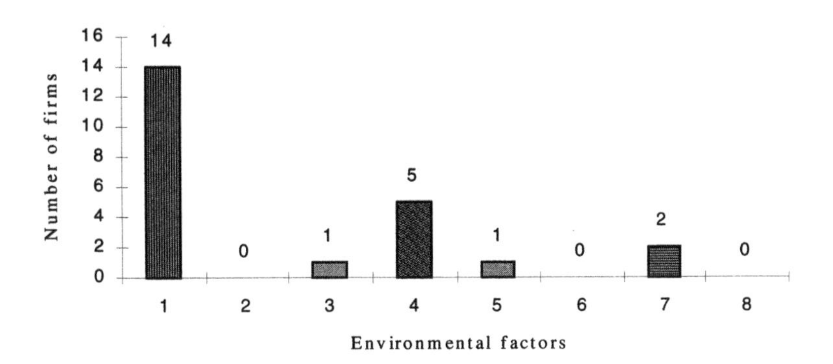

Where:
1. Government regulations. 2. Requirements of national clients.
3. Clients abroad. 4. Pressure from the local community.
5. Pressures from associations or chambers of commerce/industry.
6. Headquarter's policy (in the case of foreign ownership).
7. Public image. 8. Other (please specify).

Figure 8.3 Factors in environmental improvement. Source: Firm survey.

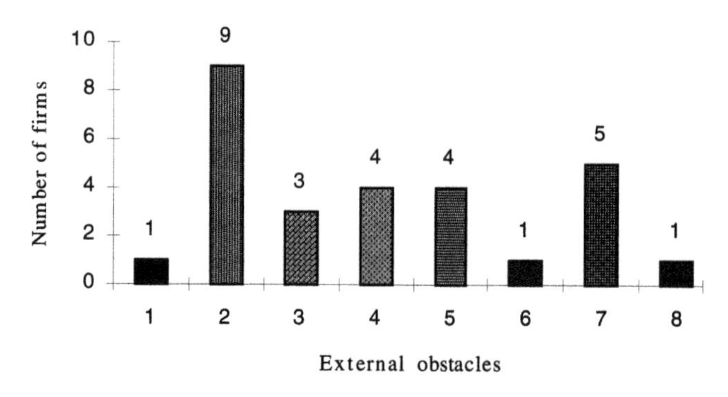

Where:
1. Lack of technology.
2. High cost of equipment required.
3. High interest rates.
4. Lack of incentives in favour of the environment.
5. Competition in the product's market.
6. Lack of local infrastucture.
7. Government policies have been an obstacle.
8. Other.

Figure 8.4 External obstacles to environmental performance. Source: Firm survey.

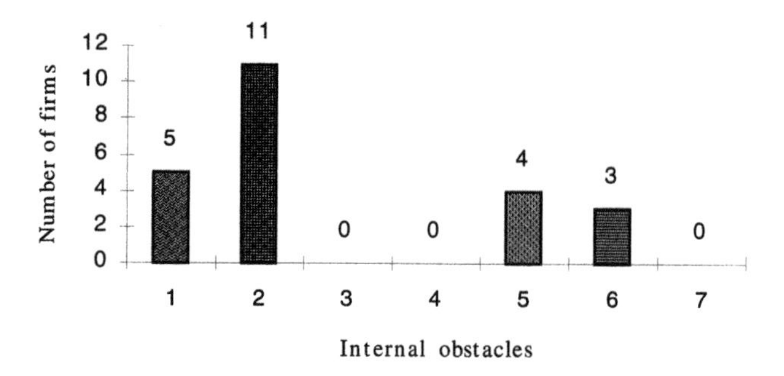

Where:
1. Inadequate funding.
2. Other priorities that require more urgent attention.
3. A lack of interest on the part of upper management.
4. Lack of skilled workers.
5. Lack of information regarding adequate technologies.
6. Inadequate knowledge regarding environmental impact and response.
7. Other (specify).

Figure 8.5 Internal obstacles to environmental performance. Source: Firm survey.

demands from local communities. They have not generally been proactive in making environmental improvements, which they see as costly in a situation where there are more pressing demands on them and where funds are scarce.

Overall environmental performance

In order to arrive at an overall evaluation of firm environmental performance, each firm's evaluation in terms of environmental policy, programmes and activities, management systems and environmental impact was aggregated. The maximum score that a firm could obtain was 46 points. On this basis the firms fell into four broad groups (Fig. 8.6).

The figure shows the heterogeneity of environmental performance in the textile industry. While a few firms have adopted a proactive stance towards environmental matters and improved their environmental performance, a significant number of firms have scarcely begun to incorporate environmental factors into their decision making.

Within the leading group of firms, three stand out because they have:

1. A written environmental policy whose principal commitment is to go beyond compliance with environmental laws and regulations and to reduce waste products.
2. Written procedures related to the effects of an environmental emergency in the plant, but not the effects of an emergency in the local community.
3. Environmental programmes to reduce liquid and solid waste discharges and to improve energy efficiency.
4. Training in environmental matters for staff in charge of the topic and for workers in general.

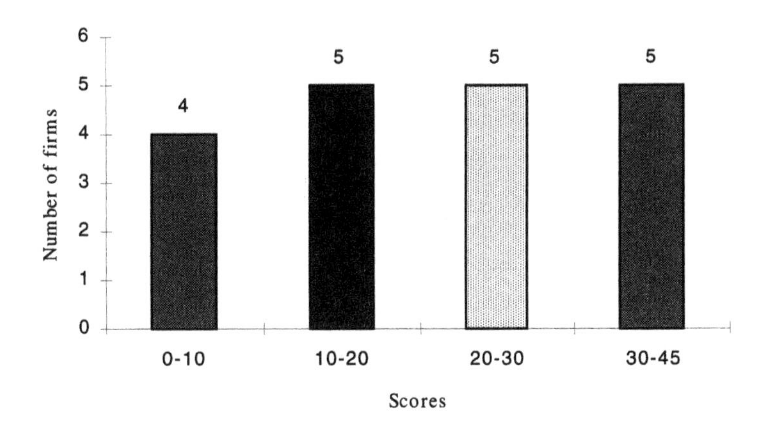

Figure 8.6 Firm scores. Source: Firm survey.

5 Voluntary environmental audits and environmental impact studies, as well as the systematic gathering of environmental information.
6 Incentives to improve the environmental performance of employees.
7 Informative publications regarding the firms' environmental management.
8 Indicators for the use of fuel, electricity and water per unit of production.
9 Samples at regular intervals and statistical series on BOD, COD and suspended solids.
10 Technology similar to the best available 2 to 5 years ago.
11 Recycling solid and liquid wastes and having facilities for the treatment of wastewater.
12 Modifying inputs, machinery, processes and products for ecological reasons.
13 Reducing the use of fuel, electricity and water.

The varied response that firms demonstrate toward environmental matters underscores the fact that they respond differently to macroeconomic policies and environmental legislation. It is, therefore, pertinent to analyse some hypotheses having to do with the environmental response of firms and the trade opening, taken up in the following section.

Trade opening and firms' environmental typology

Research carried out in several countries has pointed out that greater international transactions and increased foreign investment have led firms to modify levels of industrial pollution. This statement is based on various hypotheses. In the following sections the validity of these hypotheses is tested by looking at different groups of firms within the sample. Our aim is to compare the environmental performance of the different segments of firms.

Pollution and exports

The requirements of the international market, based in many cases on meeting the norms established by ISO 14000, are forcing export firms continuously to adopt cleaner technologies (Bridsall and Wheeler, 1992). In other words, we might hope that export-oriented firms would be more inclined to fulfil the norms mandated by the government and to incorporate technologies that pollute less than those oriented toward the home market, owing to differentiated access to technological choices and market incentives. In the case of the textile industry, we need to discern the extent to which the environmental performance of export firms differs from that of non-exporting firms.

Table 8.4 synthesises the environmental evaluation of firms classified by export levels. A number of different aspects of environmental performance were evaluated. First responses related to firm-level performance were considered, which covered the extent to which the firm had a comprehensive environmental policy, the programmes and activities undertaken, and the level of environmental management within the firm. These were then aggregated to give a corporate score. Similarly, separate scores were calculated at the plant level for environmental management, programmes and environmental impact, and these were then aggregated to give a score for environmental performance at the plant level. Finally, the firm and plant scores were aggregated to provide an overall indicator of environmental performance. Each group was evaluated, and the percentage obtained by distinct groups of firms was calculated with respect to the total possible points in all aspects of the survey. We note that export firms have a better environmental evaluation and that the differences are even higher at a corporate level.

To evaluate if these differences are statistically significant, a 'T' test, or a means difference test, was carried out. The point of the test is to compare the differences of the means obtained in each group of the environmental evaluation of export firms, with the means of the three remaining groups of firms in each aspect of the survey. Thus, the null hypothesis is that no statistically significant differences exist between the environmental evaluation of export firms and the remaining firms. Since the probability is less than 0.05, this

Table 8.4 Exports and environmental performance

	Export percentage with respect to total sales			
	+ 25% (8 firms)	*5–25% (4 firms)*	*0–5% (2 firms)*	*0 (5 firms)*
Corporate				
Policy	80	30	22	24
Programmes	70	33	17	24
Environmental management	60	21	11	19
Corporate score	61	27	15	22
Plant				
Management	78	83	56	38
Programmes	62	42	50	37
Environmental impact	30	72	44	11
Plant score	54	69	50	28
Total scores	58	48	33	25

Source: Firm survey.

hypothesis is rejected and we therefore conclude that significant differences exist between both groups of firms.

Test results show that export firms have an environmental performance better than that of the remaining firms, mainly at the corporate level. At the plant level no important differences exist, since all firms comply with environmental norms and have begun efforts at the plant level. Nonetheless, as Table 8.5 suggests, it is the export firms that have advanced in the more complex aspects of environmental performance, for example, the establishment of environmental policies and management systems at a corporate level. With regard to the establishment of programmes at the corporate level, no significant differences exist among export firms and those with export levels between 0 and 5 per cent. This group of firms has undertaken efforts to implement programmes at the corporate level.

Therefore, the empirical evidence presented in Table 8.5 would seem to confirm the hypothesis of an environmental response that differs among export firms and the remaining firms, as shown by the 'T' test for the evaluation at the corporate level and for the firm in general. While export firms have a proactive environmental response, other firms limit themselves to the fulfilment of norms.

These results also suggest that some firms with low levels of exports may be more vulnerable to the problems experienced by the sector such as the high cost of equipment, inadequate finance and competitive pressures. This indirect impact of liberalisation is an obstacle to improving their environmental performance, which leads to them seeking only to comply with environmental norms.

Table 8.5 Statistically significant differences among firms with different export levels (results of the 'T' test[a])

	5–25%	*5–0%*	*0%*
Corporate			
Policy	0.05	0.00	0.03
Programmes	0.03	0.12	0.01
Environmental management	0.02	0.00	0.02
Corporate score	0.03	0.00	0.02
Plant			
Management	0.69	0.49	0.03
Programmes	0.20	0.59	0.16
Environmental impact	0.14	0.81	0.26
Plant scores	0.33	0.94	0.08
Total scores	0.52	0.37	0.02

Source: Firm survey.

Note

a Probability that there is no difference between the environmental performance of firms in the group indicated and firms which export more than 25% of their production.

Pollution and technology

Empirical evidence from some countries (Wheeler and Martín, 1992) shows that trade liberalisation has allowed firms to obtain new technology and, therefore, to reduce pollution levels. To analyse this hypothesis within the textile industry, firms were classified according to the age of their technology. As seen in Table 8.6, the firms that have relatively modern technologies, between 2 and 10 years old, scored highest in the environmental evaluation. Scores are greater at the corporate level.

To evaluate if these differences are statistically significant, the 'T' test was again carried out, comparing the group with the most recent technology, between 2 and 5 years old, with the other groups of firms. Test results show that significant differences exist at the corporate level and, in terms of the total evaluation, between the group with modern technology and firms that have technology older than 20 years. Further, no significant differences exist in these aspects among firms that have technology between 5 and 10 years old (Table 8.7).

At the plant level, no significant differences were observed between the different groups of firms. The data in Table 8.7 allow us to state that the firms with technology between 2 and 10 years old have a proactive environmental response, firms with technology older than 20 years only comply with the norms, and those with technology between 10 to 20 years old have begun changes towards more complex activities at a corporate level in terms of

Table 8.6 Technology and environmental performance (in %)

	Technology age			
	2–5 years (8 firms)	*5–10 years (4 firms)*	*10–20 years (2 firms)*	*More than 20 years (5 firms)*
Corporate				
Policy	70	72	15	15
Programmes	53	67	44	24
Environmental management	47	64	11	15
Corporate score	54	67	22	18
Plant				
Management	74	80	56	61
Programmes	60	62	42	33
Environmental impact	52	13	44	28
Plant score	63	50	48	42
Total scores	58	58	35	30

Source: Firm survey.

Table 8.7 Statistically significant differences among firms with different technological levels (results of the 'T' test[a])

	5 to 10 years	*10 to 20 years*	*More than 20 years*
Corporate			
Policy	0.89	0.00	0.02
Programmes	0.59	0.31	0.01
Environmental management	0.54	0.01	0.01
Corporate Score	0.62	0.00	0.01
Plant			
Management	0.74	0.54	0.41
Programmes	0.93	0.19	0.01
Environmental impact	0.03	0.88	0.41
Plant score	0.44	0.68	0.23
Total scores	0.99	0.33	0.04

Source: Firm survey.

Note

a Probability that there is no difference between the environmental performance of firms in the group indicated and firms with technology between 2 and 5 years old.

environmental policies and management. Interestingly, the differences are not statistically significant at the plant level, which suggests that all firms have concentrated on improving their environmental performance at this level.

Multinational firms and environmental response

One of the consequences expected of the trade opening is greater participation of multinational corporations in local production abroad. Generally these corporations have higher environmental standards compared to national firms (Gladwin, 1987). Thus, it is expected that the subsidiaries of the multinational corporations would have an environmental response different from that of national firms in host countries. It was not possible to verify this hypothesis for the textile industry due to the fact that, in our sample, only two firms are subsidiaries of multinational corporations.

Size of the firms and environmental response

Changes in foreign trade policy have probably modified firms' strategies in various ways. Both the expansion of demand through exports as well as the recovery of the internal market forced firms to widen their production scales, reducing unit production costs by incorporating new technology and reducing X inefficiencies. It is, therefore, possible to expect a different environmental response among larger firms than the rest.

To verify this hypothesis, we compared the environmental evaluation of the largest firms (in terms of sales) with the remaining groups of firms. The data in Table 8.8 show that the larger firms perform better in the environmental evaluation. Differences are even greater at the corporate level.

As before, the 'T' test shows that the differences between the segment of firms with highest sales and the remaining firms are statistically significant at the corporate level, and at the plant level in the smallest groups (Table 8.9). These results seems to confirm the hypothesis that large firms have reduced their unit production costs by incorporating new technology and reducing X inefficiencies thereby improving their environmental performance.

Conclusions

The drop in production within the textile industry since the trade opening has led firms to decide against improving their environmental response on their own initiative. Therefore, the most important factors leading firms to take measures to improve their environmental performance have been government regulation and, second, pressure from local communities. We can, therefore, assert that the firms are facing difficulties making investments that would permit them to face the adverse impacts that their production processes have on the environment.

Table 8.8 Environmental performance by size, in terms of sales (%)

	More than 1000 million pesos (5 firms)	*100 to 1000 million pesos (4 firms)*	*10 to 100 million pesos (4 firms)*	*1 to 10 million pesos (4 firms)*
Corporate				
Policy	94	50	34	5
Programmes	79	56	27	9
Environmental management	82	28	25	7
Corporate score	83	41	27	7
Plant				
Management	84	74	56	17
Programmes	80	50	28	17
Environmental impact	40	56	6	0
Plant score	67	61	30	10
Total scores	75	51	29	9

Source: Firm survey.

Table 8.9 Statistically significant differences among firms by size (results of the 'T' test[a])

	100 to 1000 million pesos	*10 to 100 million pesos*	*1 to 10 million pesos*
Corporate			
Policy	0.10	0.06	0.00
Programmes	0.11	0.04	0.01
Environmental management	0.01	0.00	0.00
Corporate score	0.02	0.01	0.00
Plant			
Management	0.67	0.15	0.06
Programmes	0.09	0.05	0.10
Environmental impact	0.62	0.12	0.08
Plant scores	0.76	0.05	0.05
Total scores	0.39	0.05	0.03

Source: Firm survey.

Note

a Probability that there is no difference between the environmental performance of firms in the group indicated and firms with sales of more than 1000 million pesos.

The information from the survey shows that there are relatively few firms that have a proactive environmental performance. The majority have undertaken important efforts at the plant level, but the process towards more complex strategies at the firm level as a whole is still incipient, and found in only a few firms.

There are statistically significant differences in the environmental performance in favour of firms that (a) have high levels of exports; (b) have introduced more recent technology; and (c) which are larger. These results would seem to confirm a set of hypotheses regarding the impact of the trade opening on firms' environmental response. The view that the requirements of the international market force export firms continuously to adopt cleaner technologies, widen their production scales and reduce X inefficiencies is valid in the case of this industry. Nonetheless, it is worth noting that these impacts are part of an incipient process in the textile industry and limited to a small number of firms.

Notes

1 The research for this paper was undertaken as part of a project funded by the Economic and Social Research Council, Global Environmental Change Programme on *Pollution Trade and Investment: Case Studies of Mexico and Malaysia* (Award No. L320253248).

2 Acervos de Capital, Banco de México.
3 This is particularly interesting since when questioned about the attitude of top management to environmental issues, the majority of firms agreed with the statement that 'they recognise environmental problems as amongst the highest corporate priorities'. This suggests that respondents may well exaggerate the extent of their enviromental commitments when asked about them in a very general way.

Bibliography

Alvarez, L. and M. González (1987) *Industria Textil, Tecnología y Trabajo*. México: UNAM-IIE.

Bridsall, N. and D. Wheeler (1992) 'Trade Policy and Industrial Pollution in Latin America: Where are the Pollution Havens?', in P. Low (ed.) *International Trade and the Environment*. Washington, DC: World Bank Discussion Papers, 159.

CNA (1994) *Diagnóstico de las Acciones de Saneamiento a Nivel Nacional: Informe Final*. Comisión Nacional de Aguas, Subdirección General de Infraestructura Hidráulica Urbana e Industrial, Gerencia de Potabilización y Tratamiento de Aguas.

Gladwin, T. (1987) 'Environment, Development and Multinational Enterprise', in C. Pearson (ed.) *Multinational Corporations, Environment, and the Third World*. Durham, NC: Duke University Press.

Gurnham, F. (1965) *Industrial Wastewater Control*. New York: Academic Press.

INEGI (1996) *La Industria Textil y del Vestido en México*. Aguascalientes: Instituto Nacional de Geografía y Estadística.

Leibenstein, H. (1966) 'Allocative Efficiency vs X-Efficiency', *American Economic Review*, 56:392–415.

Low, P. (ed.) (1992) *International Trade and the Environment*. World Bank Discussion Papers, 159.

Mercado, A. (1985) *Estructura y dinamismo del Mercado de Tecnología Industrial en México. Los casos del poliéster, los productos textiles y el vestido*. México: Colegio de México.

OECD (1995) *Report on Trade and Environment to the OECD Council at Ministerial Level*, OED, Environment Directorate, Trade Directorate, COM/ENV/TD (95)48/ Final.

Padilla, C.(1994) 'La competitividad de la industria textil', in F. Clavijo, and J. Casar (eds) *La Industria Mexicana en el Mercado Mundial*. México: FCE Lectura No. 80.

Portos, I., (1992) *Pasado y Presente de la Industria Textil en México*. México: UNAM-IIE, Nuestro Tiempo.

Sorsa, P. (1994) *Competitiveness and Environmental Standard: Some Exploratory Results*. World Bank, Policy Research Working Paper 1249.

Wheeler, D. and P. Martín (1992) 'Prices, Policies and the International Diffusion of Clean Technology: The Case of Wood Pulp Production', in P. Low (ed.) *International Trade and the Environment*. Washington, DC: World Bank Discussion Papers, 159.

9

ENVIRONMENTAL PERFORMANCE IN THE MEXICAN CHEMICAL FIBRES INDUSTRY IN THE CONTEXT OF AN OPEN MARKET[1]

Lilia Domínguez-Villalobos

The objective of this chapter is to take an in-depth look at the way Mexican chemical fibre firms are responding to the need to control pollution and care for the environment. It is based on interviews with a sample of firms surveyed in 1997.[2] In order to examine the firm-level determinants of environmental performance in semi-industrialised nations, we explore the possible relationships between equity ownership, export activity and the size of the firm, in particular how far firms have advanced in the definition of environmental policy, management systems, programmes and performance indicators.

In addition, it is important to take into account other factors influencing the decisions of the chemical fibres firms on pollution control and pushing them towards a more careful approach towards the environment such as the strong presence of the authorities in charge of environmental matters in the chemical sector, the regional characteristics of the industrial sites of these firms and the industry's particular industrial organisation and economic performance. Our results confirm those of previous research on manufacturing firms in Mexico (Lexington, 1996; Dasgupta *et al.*, 1997; Domínguez, 1999). We shall show that, on the whole, this industry has taken steps to comply with Mexican environmental regulations. Most firms have programmes for pollution control and environmental management systems and may have taken other actions in this regard (advanced policy definition). Only a minority can be classified as non-compliant. The profile of firms which are above average in terms of environmental performance (environmental leaders in terms of the Lexington study) corresponds to that of the firms in our industry: large firms, belonging either to Mexican-owned industrial groups or large multinational corporations, with an above average level of industrial culture and satisfactory economic performance in terms of sales growth and utilisation of capacity.

The comparative examination of firms within the industry raises some points of interest. It was observed that the degree of complexity of the approach towards the environment increased with size. Multi-plant firms have explicit policies and more complete environmental management systems, but the difference is smaller in relation to pollution-control programmes and performance indicators. We found a positive association between size and capital ownership and the different aspects of environmental behaviour; nevertheless, while all firms have an above average export to sales ratio, we did not find a clear-cut association between the export-to-sales ratio and environmental behaviour of firms.

In the next section the economic background of the chemical fibres industry is presented. This is followed by an overview of the environmental problems of the industry. The following section analyses the main factors explaining the environmental performance of the chemical fibres industry. The fifth section presents in detail the results concerning the firms' responses to the need to control pollution and care for the environment. The relationship between trade and environmental performance is analysed in the penultimate section. Finally, the conclusions are presented in the last section.

Economic background

The development of the chemical fibre industry in Mexico can be summarised in three stages: the period of 'stabilising development' during the 1960s, when this industry grew at annual rates of 24 per cent by taking advantage of the protection existing within the domestic market. The second stage came during the 1970s, largely pushed by Mexico's oil boom which ended in 1981; during this period the supply and demand of chemical fibres grew at average annual rates of 10.2 and 10.9 per cent, respectively.[3] The last stage begins with the debt crisis (1981 and 1982), characterised by the stagnation of the domestic market, together with the elimination of prior import permits and the reduction of tariffs during the mid-1980s, which together dampened expansion until the early 1990s.

The basic raw materials to make chemical fibres come from the chemical industry and from forestry sources. The main source for synthetic fibres has been the public firm Petróleos Mexicanos – PEMEX – (in charge of exploration, production and crude oil refining, in addition to basic petrochemicals and some intermediate products). This has meant both benefits and limitations for the development of the chemical fibres industry. One benefit was the competitive advantage that existed due to the availability of inexpensive inputs, which were estimated to be 40 per cent of total costs (Quintana, 1997), but this also limited development. The supply of inputs from PEMEX, as defined before 1986 by the Law for Development of the Petrochemical Industry, discouraged the modernisation of plant and equipment, since the industry could count on a guaranteed supply of raw materials in very favourable conditions. After 1986, when PEMEX changed its policies and set prices to match those abroad and permitted the import of primary products, the chemical industry lost its competitive advantage in world markets, and was forced to undergo a restructuring programme that is still going on.

The supply of chemical fibres is mainly short polyester fibre, acrylic fibre and textile filament polyester, which together account for more than 70 per cent of total production. On the demand side, the principal fibres are short rayon fibre, polypropylene and polyester fibres that together account for 86 per cent of the domestic market for chemical fibres.

Industrial organisation

The chemical fibres industry is characterised by economies of scale and has been concentrated since its earliest stages. In 1979 it was made up of eight firms that operated thirty plants at fourteen locations. That same year two firms had a 64 per cent share of production, evidently a highly concentrated sector. In 1995 the chemical fibres industry had nine firms (twelve plants), specialising in the production of synthetic fibres for woven textiles. (Two of the firms, five plants, had foreign equity.) These firms meet the needs of the yarn and woven goods plants which consume almost 500,000 tonnes of natural, artificial and synthetic fibres a year.[4] The share of the four largest establishments in the chemical fibres market increased steadily from 61 per cent in 1970 to 85.6 per cent in 1988. The concentration in individual markets is also high; following analysts of this industry, the largest firms specialise in different products: Celanese is the sole producer of acetate, cellulose and rayon derivatives; only Nylon de México produces short nylon fibre; and Polifil is the sole producer of polypropylene (Quintana, 1997). Finally, some firms are part of, or have relationships with, highly diversified industrial groups or are vertically integrated.

Economic performance of the industry

The industry has had its ups and downs in recent years. After the debt crisis in 1982, it faced a drop in domestic demand, but gained ground again in 1988, when there was a tendency for domestic consumption of chemical fibres to increase until 1992, when it once more fell. Thus, between 1988 and 1993 chemical fibres remained virtually stagnant, as opposed to what occurred with GDP between 1988 and 1994. Trade liberalisation in 1987 plus overvaluation of the currency after 1991 resulted in increasing imports.

In 1995, as a consequence of the economic crisis that struck at the end of 1994, manufacturing production fell by 4.8 per cent and the chemical sector

Table 9.1 Gross domestic product (thousand 1993 pesos)

	1988	*1989*	*1990*	*1991*	*1992*	*1993*	*1994*	*1995*	*1996*
National GDP	1,042,066	1,085,815	1,140,848	1,140,848	1,189,017	1,256,196	1,311,661	1,230,993	1,190,075
Manufacturing	178,416	192,501	205,525	212,578	221,427	219,934	228,892	217,839	241,386
Chemical industry	30,418	33,279	34,725	35,060	35,684	35,075	36,270	35,954	38,296
Group: 3711 chemical fibres	1,553	1,535	1,572	1,639	1,688	1,596	1,783	2,025	2,260

Source: National Institute of Statistics and Geography (1998), *National Accounts*.

as a whole fell by 0.9 per cent, but chemical fibres (group 3711) grew by more than 10 per cent in that year and continued to do so in the next years (Table 9.1). The fact is that synthetic fibre firms have been able to compensate for the dramatic decline in the domestic market through the dynamic growth of exports, as witnessed by a greater use of installed capacity, thanks to the impact of a more than two-fold devaluation of the currency on the industry's exports (Table 9.2).

The increased exports from the industry after 1994 were made possible by the dynamic behaviour of gross investment (both foreign and private domestic investment) in spite of an unfavourable international and domestic economic situation during the years from 1988 to 1994. Gross investment grew at an average annual rate of 23.7 per cent, increasing installed capacity during the period by 20 per cent.[5]

Table 9.2: Utilisation of installed capacity in the chemical fibres industry, 1987–97 (thousand tonnes)

	Installed capacity	*Production*	*Utilisation of installed capacity (%)*	*Apparent consumption*	*Imports*	*Exports*
1987	462,000	362,333	78	280,124	7,882	
1988	462,700	365,733	79	301,295	13,136	78,574
1989	462,700	367,244	79	314,797	19,856	72,313
1990	469,733	371,873	79	334,598	30,128	67,405
1991	464,021	397,516	86	348,362	28,616	77,770
1992	470,225	431,965	92	374,613	42,987	100,339
1993	479,700	393,518	82	315,228	36,235	114,525
1994	532,020	474,369	89	328,327	42,076	188,116
1995	556,929	516,565	92	319,395	37,006	235,176
1996	569,886	580,464	102	411,068	44,439	253,345
1997	656,710	626,684	95	478,539	86,426	225,608

Source: ANIQ (various years).

Although it is known that no research and development occurs in these firms, there is an engineering development and technological learning approach, as well as an industrial culture, aspects that play an important role in making decisions regarding the environment. Productivity figures indicate that the value added per worker has increased and, notwithstanding negative growth in some years (1984–9), productivity showed favourable growth throughout the decade (5.7 per cent), above the average rate in manufacturing between 1984 and 1994. Total factor productivity had a smaller increase (1.4 per cent) owing to the high rate of investment (Brown and Domínguez, 1998).

Foreign trade

While the chemical industry in its entirety turned in a trade deficit throughout the period of this study, especially following the trade opening, and while imports of chemical fibres themselves increased at a rapid rate, this industry has enjoyed a surplus since 1991. The surplus jumped from less than US$173 million on that date to US$975 million in 1995 and US$339 million in 1996 due, in part, to the fact that imports have been held at a relatively low level, partly reflecting the contraction of the Mexican market (see Table 9.3). However, the key factor has undoubtedly been the brisk increase in exports.

The greatest volume of trade occurs with North America, origin of 74 per cent of the country's imports, and destination of 59.5 per cent of its exports. Europe is second, accounting for 5.3 per cent of Mexico's imports and 4.7 per cent of its exports. Asia is in third place, with 2.9 per cent of the country's imports, while receiving 1.1 per cent of its exports. Mexico receives from the rest of Latin America 0.9 per cent of its imports and sends there 17 per cent of its exports.

Table 9.3 International trade of chemical fibres (US$ million)

	Imports	*Exports*	*Trade balance*
1989	73.7	1.6	−72.1
1990	82.3	1.7	−80.6
1991	87.3	260.7	173.3
1992	85.4	275.2	189.8
1993	80.2	261.2	181.0
1994	118.2	655.2	537.0
1995	98.6	1074.5	975.8
1996	168.7	507.4	338.7

Source: ANIQ Yearbooks (various years).

Environmental problems in the chemical fibres industry

There are two major pollution sources linked to this industry:

1 Emissions generated by various combustion processes required by petrochemical production (boilers, internal combustion engines, turbines, etc.), producing sulphur oxide, carbon monoxide, hydrocarbons and nitrogen oxide. The emissions from some of the processes contain: (dried) ethylglycol polyterephthalate, 3.5 kg/tonne of oil steams in polyester production; (dried) polycaprolactam, 3.5 kg/tonne of oil steams in nylon production.
2 Emissions from the chemical process themselves. Some of these emissions are toxic or are health risks. They may also be inflammable and explosive, such as liquefied gas or similar compounds, combustible liquids, colouring agents, etc.

Emissions are released into the atmosphere through chimneys, but with petrochemical processes there may also be numerous leaks from storage tanks, and from joints and valves, which depend a great deal on the characteristics of the operation and maintenance of the installations.

Geographical location has much to do with the impact on the environment. For example, facilities located on the coast have favourable weather conditions with air currents that disperse pollutants, preventing their excessive accumulation. On the other hand, facilities in the centre of the country have unfavourable weather conditions for dispersion of pollutants.

The petrochemical industry consumes large quantities of water. In fact the chemical industry is considered to be one of the nine most important in terms of water consumption. In the chemical fibres sector, consumption varies between 0.3 and 30 m^3/tonne. Water consumption occurs in two processes (to obtain caprolactam and ethylglycols): first, that which mixes the products to be made, or when it is disposed of, contaminated with unwanted by-products; second, the water used in cooling, where part is evaporated and is disposed of with a high salt content resulting from the purging of the cooling towers (typically in the process to obtain ethylglycols). The disposal of waste is a serious problem for water pollution, because in Mexico some firms discharge their wastes into rivers, lakes and/or the sea.

The principal pollutants are organic substances that increase the demand for oxygen within the effluents; among these products are those such as grease and oil that are not water-soluble and thus float. The most dangerous pollutants, which demand greater attention, are the cyanides, heavy metals from catalysers and diverse organic chlorides.

The petrochemical industry is one of the sectors that poses greatest risk to the environment, because of the characteristics of the products used, and due also to the complexity of the processes. The use of flammable, toxic, explosive and corrosive products is common in the industry, as are high temperatures and pressure. Security valves, cut-off valves, back-up systems for electricity, refrigeration, etc., are some of the security measures normally included in the plants, but aspects of accident prevention and emergency procedures require that more attention be given to the environmental impact of these installations.

Interviewees recognised facing various types of pollution: air, water, solid wastes and hazardous waste. Plant discharges go into the industrial drainage system (four plants), a river (one plant), a lake (one plant), a dammed reservoir (one plant) and in one case it is used for watering greenery at the complex. Water treatment is undertaken at the plant or through the municipal system. The percentage of hazardous waste is approximately 40 per cent of the total generated.

Factors that explain the environmental performance of the chemical fibres industry

Decisions having to do with the care of the environment in this industry cannot be understood simply by the differences among firms. Among the most important factors are, notably, the restrictions arising from environmental norms and the involvement of the environmental authorities in the different branches of the chemical industry and the geographic location of the firms where there is a high degree of environmental deterioration. In what follows we briefly present the way in which these factors have an impact on the decisions of firms in the chemical fibres industry. We then review the characteristics of the firms and the results of the questionnaire, answered by all plants of the six firms, concerning the motives and obstacles affecting the environmental decisions taken at the plant and corporate levels.

Involvement of the environmental authorities in the chemical sector

Worldwide, the chemical industry is considered a high-risk sector for the environment. The constant attention shown by environmental authorities is due to the characteristics of the chemical fibres industry and the potential problems that can arise if left unchecked. Recently, significant progress has been made regarding emission control, and disposal and management of hazardous waste. Likewise, there is a tendency for water fees to increase based on levels of consumption and discharges. Efforts made by the authorities to enforce environmental regulation have been stepped up among high-risk industries (those which are fuel intensive and highly polluting). These receive a greater number of visits, and firms have been encouraged to participate in voluntary audits.

According to the firms, this attention has been particularly intense since 1992 when an explosion destroyed an area 2 km long in Guadalajara, Jalisco.

The chemical branch comes under federal jurisdiction and is therefore subject to overview by the Federal Environmental Agency (PROFEPA). Because they handle hazardous wastes and explosives, chemical fibre firms are required to undertake risk assessments, to declare the transport and storage of hazardous waste and to have an inventory of their emissions. They are also required to make environmental impact assessments when opening a new line of production, and must comply with water, emissions and hazardous waste norms. PROFEPA's responsibilities include verifying that sources of pollutants within federal jurisdiction comply with environmental rules, whether they derive from laws, regulations, norms, authorisations from the National Institute of Ecology, or from the delegations of the Secretary for the Environment, Natural Resources and Fishing (SEMARNAP). This latter institution undertakes inspection visits and, after analysing samples, issues resolutions. A programme is then drawn up with measures to be applied immediately and over longer periods. It is PROFEPA's responsibility to verify that measures are being implemented.

PROFEPA reported having undertaken 47,096 visits at a national level between 1992 and 1996. The results of these visits were as follows: 70 per cent showed minor irregularities; 4 per cent led to partial shutdowns; 0.008 per cent led to total shutdowns; and 25 per cent turned up no irregularities. The sectors most often visited by PROFEPA are the chemical–petrochemicals, metal mechanics and pulp and paper industries.

The firms' responses confirm the importance of environmental regulations. Five of the six firms interviewed mentioned compliance with laws as the most important factor in their environmental decisions, and the sixth placed it in second place (pressure from the community was first). Nine of the ten plants reported having participated in voluntary audits within the 'Clean Industry' programme; these audits outline programmes to be implemented and target completion dates. Firms are required to purchase a bond which is returned once the programmes have been implemented and a Clean Industry certificate has been issued. We conclude that governmental action has been an important factor for all of the firms.

PROFEPA data confirm that in the chemical fibres industry one of the plants has concluded its programmes and has been certified, two plants have fulfilled their action plans and are about to be certified, in three cases the action programme is underway, in two cases it is being discussed and in one plant the audit is underway.

The regional factor

The chemical fibre firms are located near the main markets, either in the large cities (three are in Monterrey and one is in Mexico City, the second and first industrial centres in the country) or in industrial corridors close to them (two of them in Jalisco next to Guadalajara the third industrial centre, one in

Puebla and another in Queretaro next to the Mexico City Metropolitan area and one in Tampico, relatively near to Monterrey). The influence of geography can be a determining factor in the firm's environmental decision in cities with one or several of the following problems: (a) lack of water, (b) rapid urbanisation, and (c) fragile ecosystems.

Lack of water

Although lack of water is a problem throughout Mexico, with the exception of the southeastern portion of the country, it is particularly serious in the cities of the state of Nuevo León. Three of the plants visited are in Monterrey, a city with relatively little and irregular rainfall, coming mostly towards the end of summer. The climate is dry, with extreme temperatures and there is little vegetation. Nuevo León's advanced industry and factors inherent to the region have led to the disappearance of underground water sources and the drying up of rivers during most of the year. A large part of the region's water comes from wells. Since there is no permanent flowing water in the region, competition for it is fierce. A similar situation exists in San Juan del Río, in the state of Querétaro, where another of the plants is located.

The demand for water is such that firms seek to rationalise its use, and most firms have been able to reduce consumption. Further, the lack of water has encouraged the use of treated sewage water within the industry. As an example, one of the firms visited has had a water treatment plant since 1956 at its Monterrey complex, the site of one of its chemical fibre plants. Water used by the firm comes from the city's sewage. It is taken directly from the drainage system and is given primary, chemical and biological treatment and then used to generate steam and electricity. It is returned to the drainage system after another primary treatment. The firm has applied the experience gained in engineering and management of water treatment plants in other plants, and this is now a source of income from its environmental services firms, as will be seen later. It is common to find treatment plants for the use of sewage water for industrial uses in Monterrey.

Rapid urbanisation

The closeness of the plant to residential areas, as a result of a lack of urban planning, has become a decisive factor in firms' decisions regarding the environment. Both Mexico City and Monterrey are the birthplace of the country's industrialisation, and grew very rapidly from the 1940s. Many people migrated there, attracted by the industrial surge, employment opportunities and the rise in the standard of living. The resulting urban sprawl in these cities was such that industries located in relatively isolated areas 20 years ago soon found themselves surrounded by residential communities, increasing the consequences of accidents as well as raising community awareness about the firm's activities. Over time this situation has been repeated in mid-sized cities in the states of Querétaro and Mexico, where chemical fibre plants visited are located. Firms studied here have been affected by urban sprawl. Significantly, community pressure is reported to be increasingly important. In two cases the community threatened to close the plants on one occasion or another. Six plants reported implementing programmes in accident prevention, to reduce emissions, odours and noise, and to mobilise in the case of an emergency, as well as community relations programmes. One firm has a 24-hour telephone hotline for complaints from the community.

Fragile ecosystems

The cities where the chemical fibre plants are located, especially Monterrey, Mexico City and San Juan del Río, Querétaro, have suffered serious environmental degradation. Water discharged in Mexico City has 513,180 tonnes of biochemical oxygen demand (BOD),[6] of which 63 per cent is estimated to be industrial (González Palomares, 1995).

The San Juan river basin in the state of Nuevo León is considered to be the second most polluted area in the country. The Monterrey metropolitan area discharges approximately 7131 litres per second. Although only 19 per cent of the discharge is industrial, the industry's share of BOD is proportionally higher.

Last, although San Juan del Río, Querétaro, is a small city which industrialised recently, important pollution problems now exist, which, according to reports from SEDUE, directly affect the water quality of the San Juan River and the Centenario dam. Nearby rivers, such as the Querétaro and Pueblito, are very polluted.

Before passing to the factors that individually affect the firms' environmental actions, it is important to note that the common complaint among industrial firms about the economic instability as an important factor delaying environmental decisions[7] is weaker in the case of chemical fibres from the overall economic performance described above. The recent economic performance of the industry has been better than the industrial average and there are no small and medium-sized firms in this industry. Thus, we should not expect these factors to have the same importance in the case of firms in this industry.

Motives and obstacles behind firms' environmental actions

It is not always possible to order strictly the factors that motivate a firm to undertake measures to improve the environment. The factors mentioned most frequently are, first, government regulation (three firms), community pressure (two firms), headquarters' policy (one firm) and the need to lower the transformation indices (one firm), meaning the aim to increase efficiency in

Table 9.4 Influence of factors on environmental performance

	First	*Second*
Environmental regulation	3 (5 plants)	1 (2 plants)
Local customers' requirements		
International customers' requirements		2 (4 plants)
Local community pressures	2 (3 plants)	
Associations or chambers'		
policy of mother company	1 (2 plants)	
Public image		2 (3 plants)
Other	1 (1 plant)	

Source: Firm survey.

the use of inputs. In second place, factors such as the requirements of international clients and public image are mentioned (see Table 9.4).

The external obstacles that firms identified as inhibiting their environmental response were government policies (three firms), the high cost of equipment (one firm) the lack of technology (one firm), and the high interest rate. It is paradoxical that while governmental regulations are mentioned as a fundamental incentive to improve the environment, they are also mentioned as obstacles. One firm mentioned the lack of co-operation by the authorities in regulating the growth of communities around the firms; another stated that government foot dragging in regulatory matters was initially similar to that of the firms. In this author's opinion these responses show unhappiness with some government agencies, but this can hardly be interpreted as a limiting factor in firms' actions to control pollution.

Regarding internal obstacles to environmental response, the most important were: other priorities (four firms), inadequate knowledge for solving environmental problems (two firms) and a lack of financing (one firm). In spite of the fact that the overall performance of the chemical fibre industry is above average, under 'other priorities' firms mentioned mainly liquidity restrictions (see Table 9.5).

Responding to the need to control pollution and care for the environment

Basic elements of firms' environmental policies

Of the six firms visited, three reported having an explicit written environmental policy, and three have no such policy. The former firms are characterised by having several plants and are, in fact, those with the greatest market share. This policy includes explicit commitments to comply and go beyond Mexican laws, and continuously improve environmental response

Table 9.5 Obstacles to environmental performance

	First	*Second*
External		
Lack of technology	1 (1 plant)	1 (3 plants)
High cost of equipment	1 (2 plants)	1 (2 plants)
High interest rates	1 (2 plants)	0
Lack of environmental incentives	0	0
Market competition	0	0
Lack of local infrastructure	0	0
Public policy	3 (4 plants)	1 (3 plants)
Other		1 (1 plant)
Internal		
Inadequate finance	1 (2 plants)	1 (3 plants)
Other priorities	4 (6 plants)	1 (2 plants)
Lack of interest from superiors	0	0
Lack of qualified labour	0	0
Insufficient technological information	0	
Inadequate knowledge	2 (4 plants)	1 (4 plants)
Other	0	0

Source: Firm survey.

and participation within the community. Two of the firms also have a commitment to sustainable development. The three firms signed a declaration at the sector level with the National Association of the Chemical Industry (ANIQ) regarding integral responsibility. This is an important declaration because it aims to change the way companies operate and to develop ways of continuously improving, leading companies to, first, comply with laws and regulations in force in Mexico, as well as maintain fruitful relationships with the authorities and surrounding communities, and, later, upgrade performance levels to improve competitiveness in domestic and international markets.[8] Another important point refers to following through on codes and administrative practices that endeavour to cover a product's complete life cycle, implying responsibility for its care from its conception to its final disposal. The seven codes of administrative practices cover the following aspects:

1 Protection of the community
2 Prevention and control of environmental pollution
3 Safety of the processes
4 Safety and health at work
5 Transportation and distribution
6 Research and development
7 Product safety.

The Administrative Practices Codes establish qualitative goals and objectives, which supposedly should be fulfilled at each stage of the product's life, meaning that every company must carefully analyse its content, adapt it to its own needs and later draw up an individual strategy for implementation. Two firms stated that they had fully participated in the creation of this programme within ANIQ.

The three firms previously mentioned consider the environment, industrial security and occupational health to be related concepts. The reason, they state, is that in the chemical industry, any flaw in the processes would directly affect workers and could affect an entire community, as in the case of the catastrophes at Bhopal or in San Juan Ixuatepec in Mexico. The three firms have explicitly declared that these concepts have been incorporated in their competitive strategy and in their commitment to business excellence.

Top management at these firms has an attitude towards the environment that has been given *the highest priority* within the corporation and this fact is communicated to employees. The companies have documents regarding the effects of an environmental emergency in the plant and of an environmental emergency in the local community, in addition to having an environmental impact assessment of new projects and procedures for the safe disposal of the plant's industrial wastes.

There is a learning factor related to environmental care, similar to the technological learning process that accompanies the import of technology. The three firms with an explicit environmental policy stated that their experience in environmental matters had been strengthened in one case by foreign partners (principally Dutch) and in the other case by their German and US headquarters. The two foreign firms (five plants visited) responded that the objectives of their headquarters' environmental policy *vis-à-vis* their subsidiaries had to do with their own environmental norms, which are more stringent than in Mexico. In the USA these firms participate in the Responsible Care programme, begun in 1985 by the Canadian Chemical Producers' Association, to which several other countries have subscribed. This is an international programme that covers performance codes for: (a) preparing for emergencies and community response; (b) health and well being of employees; (c) the safe distribution and transportation of raw materials and products; and (d) safety in processes and in products, as well as pollution control. Head office monitors the firms in biannual meetings in one case, and with protocols similar to those found in ISO 14000 every two years in the other. Head offices supply research and development services in order to help solve the subsidiary's environmental problems. On the engineering side, a regional corporate group exists, in charge of overseeing projects, construction and maintenance, independently of whether the plants have their own engineering staff (see Table 9.6).

The firms that reported not having an explicit policy, even though they are not small firms, have a relatively low share in the sector. Each has only one

Table 9.6 Environmental policy

Does the firm have an environmental policy?	*Yes*	*No*
Explicit policy	3	3
Signature of formal commitment or a document of environmental principles at the sector level	4	2
Recognition of environmental issues among the highest priorities of CEO	4	2

Source: Firm survey.

plant, which can in part explain their less systematic approach to environmental problems. This does not mean, as will be seen below, that compliance is not important. For example, one firm within this group, located in Mexico City, with strong demands from the surrounding community and from environmental authorities, participated in ANIQ, and its top management has placed the highest priority on responding to environmental problems. It is relatively advanced in terms of its procedures in environmental management as compared with other firms.

Environmental management

As in the case of policy, we observed that the larger firms tend to have a formal environmental management system at a corporate level, including a manual, while firms with only one plant had no similar system. There is someone in charge of environmental matters in all firms, but only within the group of larger firms was the post exclusively for that purpose. These larger firms also publish the results of internal environmental improvements and work with the firms' internal community. Two firms have introduced environmental concerns in their incentive systems. Other firms considered they were part of performance incentives based on quality circles. We were able to observe that two firms grant recognition, through an internal magazine, to the best environmental project or to the group of workers with the lowest rate of incapacitating accidents (see Table 9.7).

Although not all firms use a formal approach, most firms have basic aspects of environmental management, such as documents on the effects of an environmental emergency, and the safe disposal of wastes, in addition to indicators of environmental performance, with short or mid-term follow up. At the plant level, most firms have indicators to measure water and energy consumption per unit of output, as well as emissions or discharges. These indicators are followed periodically in order to reduce waste. In some firms these indicators have only recently been created (see Table 9.8).

Table 9.7 Environmental management

Measures of environmental management at the firm level	No. of firms
Specific post for environmental matters	6
Manual of environmental management	3
Performance indicators	5
Collection of information about environmental performance	5
Publish improvements in journal or newspaper	3
Uses incentives related to environmental performance	2

Source: Firm survey.

Programmes to improve environmental performance

Figure 9.1 shows the responses of firms in relation to specific programmes to improve environmental performance. All the plants visited have programmes to reduce pollution from different sources, in contrast to formal policies and management systems, which not all plants have.

Except for one plant, all have programmes to reduce energy consumption. Firms have always claimed that electricity prices are higher than those of their competitors. Moreover, they complain about interruption of electricity supplies which may imply an additional cost over electricity prices (Máttar,1994). Here, it is worthwhile noting that one firm has an ambitious project to be completed in two years for the co-generation of electricity via a 120-MW plant at Altamira, Tamaulipas. The project will cost some US$75 million, with a thermal capacity of one million pounds of steam per hour. The plant will replace energy sources at other plants in the complex where raw materials for fibres are made. Electricity will be supplied to firms in Monterrey through a transmission contract with the CFE (Federal Electricity Commission). It is estimated that joint savings in energy at this plant will be equivalent to one million barrels of oil per year.[9]

Under 'other', the largest firms mentioned programmes for safe transportation of raw materials and hazardous waste, recycling or programmes for soil

Table 9.8 Aspects under environmental management

	Plant
Emissions indices	9
Water consumption per unit of output	8
Discharge indices	8
Energy consumption per unit of output	6
Fuel consumption per unit of output	8
Monitoring at regular intervals	9

Source: Firm survey.

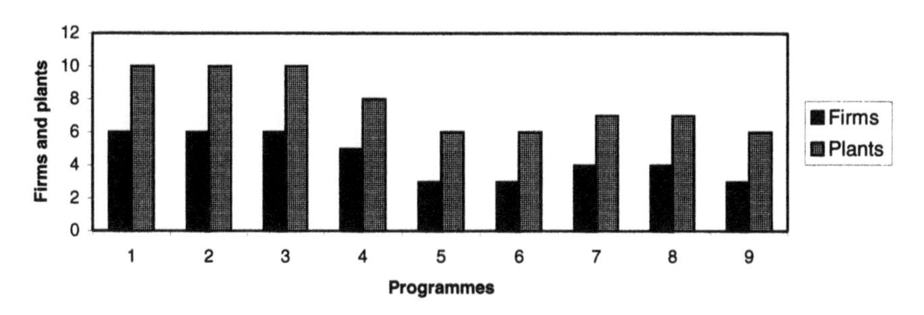

where:
1. To reduce air emissions
2. To reduce discharges(quantity and/or quality)
3. To reduce solid waste
4. To increase energy efficiency
5. Other programmes such as soil redemption
6. Quantitative goals
7. Training programmes
8. Participation in voluntary audits
9. Programmes involving local community

Figure 9.1 Programmes and activities related to the environment.

remediation. One of these firms mentioned a US$25 million joint venture project for a recycling centre for hazardous waste in Coatzacoalcos, Veracruz, which will service the firm and others in the area in addition to extracting synthesis gas, i.e. a raw material. The technology supplier will install the reactor and the firm will grant land and infrastructure. It is anticipated that the project will be finished in one year.

Five of the companies visited have training programmes related to environmental matters, which include courses for environmental staff and for production workers. In addition, in one of the firms all personnel receive a three-day induction course on environmental matters upon starting employment. Most firms include topics related to waste minimisation, water treatment, environmental legislation and air pollution control in their training programmes. The largest firms (four plants) cover other topics such as risk analysis, environmental management and control, environmental audits, environmental impact analysis and ISO 14000. It was not possible to obtain comparative figures on worker hours in environmental training.

As mentioned previously, five of the six firms have undergone or are presently undertaking an environmental audit, some having been obliged to do so by the government. They also participate in voluntary audits undertaken by the environmental attorney's office (PROFEPA) out of which they defined a whole set of actions to be corrected in a period of time. They also report having done environmental impact studies for new production lines, in addition to detailed industrial risk analysis. Four firms have community outreach programmes on environmental matters. These programmes call for emergency brigades staffed by community volunteers to plan for emergencies and review accident prevention, as well as to carry out practice drills and simulations to verify their effectiveness.

In short, pollution control and training programmes are present in most firms within the industry. However, larger firms tend to have a more systematic approach, suggesting that environmental care is not simply a matter of resolving problems in the short run, but needs a long-term strategy including quantitative goals. We found deadlines to fulfil these goals run from 3 to 5 years in three cases and between 1 and 2 years in the other.

When questioned about the type of changes introduced for ecological reasons, equipment related changes were the most important, followed by process changes and changes in inputs. Only one plant reported a product-related change.

Changing equipment to improve the ecology has various objectives: to collect different types of effluents (two cases), installation of a biological or chemical water-treatment plant (four cases), a system to recycle residual water (two cases), gas scrubbing equipment (one case), continuous monitoring of combustion gases (one case) and a microwave cleaning system (one case).

Changing the production process for ecological reasons include avoiding discharges (two cases), reducing raw materials waste (three cases) and making combustion more efficient. Lastly, with regard to changes in inputs for environmental reasons, the most frequent action was the change of fuel in favour of natural gas (four cases), substitution of chromate by borate (one case), changing additives (one case) and substituting demineralised water for distilled water in yarn production.

In our interviews, emphasis was placed on the fact that some ecological measures can help the plant use its resources more efficiently by encouraging savings in fuel or raw materials. For example, one of the larger firms implemented a project to reduce waste in polymerisation and yarn making at one of its plants, a project that shows how a continuously improving process implies bringing together quality, safety and environmental improvement, in addition to improving worker and community living standards, while increasing business' productivity, generating 625.8 tonnes less waste with savings of 3 million pesos.

Other changes are aimed at waste recycling programmes. In two cases petroleum wastes are sent to a cement firm as an alternative fuel. Others are recycled or treated at the plant. One of the firms took plaster produced from one process, raised it to industrial quality by means of a joint agreement with a construction firm, and now its disposal is cost free.

The case of a product change for environmental reasons involved closing one product line, since it required an electricity generating plant which used fuel oil extensively and was a heavy polluter. By closing the plant and substituting electricity supplied by the electricity company, fuel consumption dropped dramatically. The decision had profound consequences, since it implied shutting down the company's chip production line, which would have been unprofitable under present circumstances, thus leaving only the extrusion and spinning processes. The firm now buys its chips from other firms. In this case it was the community that originally applied pressure in 1992 and the firm decided to close a line of production rather than move to another site.

Environmental impact

Before reviewing aspects related to environmental performance, the firms' opinion regarding the sources of environmental improvement are useful for understanding the philosophy behind their acts: systematic management could be interpreted as an attempt to control pollution in a more integral manner; the modification of processes attempts to correct problems at the source; and treatment can be an effective solution to an urgent and short-term problem. Three firms consider the first source of environmental improvement to be environmental management systems and two firms mentioned the modifications of processes, while one firm mentioned the treatment of emissions and discharges. Recycling was cited by three firms as the second most important contribution, and three mentioned treatment. Modifications in the products were not considered a significant source of environmental improvement, although stopping the production of a particular item was important in one case.

The companies' self-evaluation of their impact varies, but in all cases there is agreement that plants are complying to domestic norms. Four firms consider their performance comparable to that of a plant abroad, or comparable to the best standards domestically in the industry, exceeding at least the legal requirements. Two firms aim to meet local environmental norms, although they sometimes failed on specific points.

The firms report that targets set as a result of environmental audits have been met by over 70 per cent in two of the companies, and between 30 and 70 per cent in the other. For example, in its report to the board of directors, one firm reported a 64 per cent drop in total emissions, as compared to 1991 baseline figures. This reduction largely occurred during the first two years. Improvements in later years have been very slow and were said to be more costly, but the downward tendency continues. Accidents related to raw materials or transport of waste have dropped by 40 per cent since 1994, and hazardous wastes have been reduced by 10 to 15 per cent, thanks to a waste reduction programme.

When asked if the firm had been found in breach of government norms, five firms responded affirmatively, but added that violations were not frequent.

The most common problems were pollution by wastewater or emissions into the atmosphere. Sanctions applied against the firms were fines and orders to fix the problems, undertaking voluntary audit programmes given by PROFEPA to pinpoint problems and establish deadlines for resolving them.

Performance at the plant level was examined by the reductions in certain indicators measured per unit of output, as well as emissions and discharges. The frequency and size of these advances can be seen in Table 9.9. This shows reductions in wastewater discharges (from 9 per cent to their virtual total elimination) and falls in BOD and COD values, as well as increased rates of recycling and reduced emissions. Important savings were also achieved in fuel use, as shown by the reduction in consumption per unit of output. It is clear that most plants have achieved important advances.

Final evaluation of the firms

Following the methodology outlined in Appendix 2, we now evaluate firms' performance regarding each aspect, with a final score to reflect the extent of the firm's efforts to reduce pollution.

In order to do this, firms were divided into three broad categories: from 75 to 100 per cent (high), from 50 to 74 per cent (medium), and less than 50 per cent (low) of the theoretical best performance. The results are presented in Table 9.10.

With the caveat that any rigid classification of firms' environmental performance should be treated with caution, we can conclude that six plants

Table 9.9 Some environmental achievements

		Reductions (%)							
Plant	*Recycling (%)*	*Fuel*	*Electricity*	*Water use*	*Water discharge*	*Emissions*	*BOD*	*COD*	*TSS*
1	0	−90	+	−50	−50	−95	[a]	−80	−30
2	20	−1	−15	−1		[a]	[a]	[a]	
3	50	−5	+5	−12	−98	−15	−79	−47	−76
4	20	−7	−4	−2	−5	−90	−10		−49
5	35	−20	−25	−40	−40		−20	−25	
6	85	+	−8.5	−15.3	−55	−50	−30		
7	40	−6.5		−9.6	−9	−30	−84	−63	[a]
8	10		−20	−1.5	+20	−25	−30	−25	[a]
9	0	[b]	[b]	[b]	[b]	[b]	[b]	[b]	[b]
10	40	+		−25	[a]	[a]			

Source: Firm survey.

Notes

a Downward trend, but no rigorous statistics at the beginning.

b Did not answer.

+ = percentage increase.

Table 9.10 Number of plants classified according to environmental achievement

	High	*Medium*	*Low*
Policy	7	1	2
Corporate environmental management	5	2	3
Plant environmental management	7	2	1
Corporate programmes	5	4	1
Plant programmes	5	5	0
Environmental impact	6	1	3
Corporate evaluation	7	1	2
Plant evaluation	6	3	1
Overall evaluation	6	2	2

Source: estimated from the firm survey.

have a clearly defined policy, a formal system of management, as well as environmental programmes in different areas, with a performance that should be considered excellent. Two plants have characteristics that indicate proactive behaviour, given their environmental programmes and the quickness of their response, which includes concluding their PROFEPA voluntary-audit action programme. One plant can be said to be complying, given its score. One of the plants has an unsatisfactory score. Thus, the majority of the plants in the industry are complying or going beyond compliance, especially since among the firms not interviewed here, there was one that had been certified under the government's 'clean industry' programme.

We pointed out at the beginning of this chapter that there are specific industry factors that partly explain the response towards the environment: environmental regulations and the intense activity of governmental agencies within the chemical branch, the restrictions imposed by their regional location in terms of lack of water, rapid urbanisation and the fragility of the ecosystems. Also included here are the economic factors and the economic performance of the industry itself, especially its high degree of technical and economic concentration, and the relatively important degree of opening to markets abroad (i.e. all firms export). In the next section, we conclude by examining the differences in the performance of firms by equity ownership, size, technology and export activity.

Understanding environmental performance in an open market

Given the transition from import substitution to export promotion, in which multinational companies play a key role in foreign trade through intra-firm trade, it is worthwhile examining the average scores of domestic firms versus those with multinational interests (taking as foreign any firm with foreign equity exceeding 40 per cent). Comparing domestic and foreign firms, we found scores are, on average, higher in the latter. As mentioned, foreign firms emphasised the interest of headquarters in having their subsidiaries not only comply with domestic standards, but also with those set abroad by the corporation itself.

In Table 9.11 one has to bear in mind that among domestic firms, some may have a proactive environmental behaviour, but their efforts are lost in the averages. Thus, perhaps an important variable to explain the differences in terms of policy formalities and environmental management has to do with the size of the firms analysed here. All the firms in our sample were placed within the upper rank in our classification for sales. We thus opted to differentiate among those belonging to the first 100 firms in the annual survey of a Mexican business magazine.[10] These firms are part of large, diversified corporations or are vertically integrated.

Table 9.12 shows that, comparing the three firms within the top 100 with the rest of the firms, the larger ones post much higher averages than the smaller ones (83.7 per cent versus 65.2 per cent).

An open market encourages processes of technological adjustment. None of our firms had taken, at the time of the interview, a radical approach of technology updating. However, they stated that modernisation is not always a matter of changing equipment, but also involves introducing continuous improvement systems and soft technology systems. Indeed, firms today seek to be more efficient in order to be more price competitive and improve quality. This may mean that savings in materials and fuel are led by both efficiency and environmental concerns. We found no linear relationship between new technology and the average score of firms,

Table 9.11 Environmental performance by capital ownership (% of maximum number of points)

	Foreign ownership 2 firms (5 plants)	*National 3 firms (4 plants)*[a]
Policy	100.0	67.0
Corporate management	100.0	59.1
Plant management	96.3	91.1
Corporate programmes	73.7	76.7
Plant programmes	80.0	66.7
Environmental impact	86.7	66.7
Plant evaluation	90.0	77.5
Total evaluation	86.5	71.7

Source: Firm survey.

Note

a Averages obtained from survey of 9 plants. Data from one plant were discarded, since evidence was insufficient.

Table 9.12 Size and environmental performance (% of maximum number of points)

	3 giant firms	*The rest*
Policy	100.0	45.0
Corporate management	90.9	18.2
Plant management	84.4	100.0
Corporate programmes	81.7	75.0
Plant programmes	78.0	67.0
Environmental impact	86.7	66.7
Plant evaluation	83.3	77.1
Overall evaluation	83.7	65.2

Source: Firm survey.
Note: Averages obtained from survey of 9 plants. Data from one plant were discarded, since evidence was insufficient.

although technologies of the oldest vintages proved to have the lowest averages.

Lastly, the hypothesis regarding the relationship between exports and environmental performance was examined with the exports-to-sales ratio from each plant. As can be seen from Table 9.13, at first glance there is an inverse relationship. This was rather surprising. It is necessary to interpret this result with caution. First, two of the large corporations have been among the 250 largest exporters for several years and yet some of their plants are in column 2, while two firms with a lower value of exports appear in column 1,

Table 9.13 Technology level and environmental performance (% of maximum number of points)

	Technology level comparable to the best level of		
	Last 5 years (2 plants)	*5–10 years ago (2 plants)*	*10–20 years ago (5 plants)*
Policy	100.0	100.0	86.0
Corporate management	86.4	100.0	90.9
Plant management	94.4	100.0	50.0
Corporate programmes	75.8	75.0	75.0
Plant programmes	90.8	83.3	70.8
Environmental impact	77.8	100.0	63.9
Plant evaluation	85.4	95.8	74.2
Overall evaluation	85.2	93.9	73.3

Source: Firm survey.
Note: Averages obtained from survey of 9 plants. Data from one plant were discarded, since evidence was insufficient.

giving the averages a downward bias. Second, export-to-sales ratios may change from year to year so that one year's information may be insufficient to analyse the relationship between trade and environmental performance.

In any case it must be recalled that environmental requirements from international clients, although mentioned in some cases, were not considered the most important motivation in environmental decisions compared to other factors such as environmental regulations and pressures from communities, which at times may be stronger for a firm exporting on a smaller scale (see Table 9.14).

Conclusion

Summing up the main findings:

1 Our results show that, on the whole, this industry has taken steps to comply with Mexican environmental regulations. Of the six firms surveyed, three can be classified as excellent, one as having a proactive response, one in compliance and one whose performance was questionable. In seven plants belonging to three firms, environmental policy was explicit and given high priority. These firms have a formal management system with complete documentation and systematic programmes. The lack of a formally defined policy did not mean that environmental matters were not important. We found that irrespective of whether the firms had an explicit policy, most firms had formal environmental systems in addition to programmes in different areas to control pollution, including modifications in inputs and production processes, and not just 'end-of-pipe' solutions.
2 Among the most important factors affecting the decisions of chemical fibre firms are the restrictions arising from environmental norms accompanied by the involvement of environmental authorities monitoring compliance in the different branches of the chemical sector in the context of a high degree

Table 9.14 Exports and environmental performance (% of maximum number of points)

	Exports > 50% (4 plants)	*50% > Exports > 25% (4 plants)*	*25% > Exports > 2% (1 plant)*
Policy	86.0	88.0	100.0
Corporate management	69.1	77.3	100.0
Plant management	83.3	93.3	88.9
Corporate programmes	79.2	73.3	100.0
Plant programmes	88.3	78.3	81.7
Environmental impact	61.1	85.6	88.9
Plant evaluation	72.1	88.5	91.7
Overall evaluation	73.9	83.9	91.1

Source: Firm survey.

of environmental degradation and the fast unplanned urbanisation of the regions where the firms are located. An interesting finding is related to the potential role of PROFEPA in this learning process. Firms reported that voluntary-audit action programmes have been a vehicle for systematising solutions to environmental problems within the firms. Our results show that firms located in high-growth urban areas also face great pressure from the community, which at times is more important than government regulation. In short, regulation matters as well as public scrutiny.

3 The chemical fibres industry has had above average output and export growth in the last four years. Thus, economic instability has affected them less than in other industries. In addition it must be remembered that the industry is highly concentrated and made up of large firms, which explains the relative homogeneity of performance in comparison with other industries where small and large firms coexist. Although liquidity restrictions were reported as a factor to be considered in environmental decisions, firms also recognised internal shortcomings in knowledge and the need to undergo a learning process.

4 The fast transition to an open economy has implied increasing multinational investment, an ongoing process of technological updating and flourishing exports in the chemical fibres industry. Our results show that relationships with foreign partners have been a contributing factor in speeding up the design of policy and the introduction of environmental management in those firms with above average environmental performance. Also, plants belonging to diversified corporations had a better environmental performance. These firms had higher average scores, which suggests the presence of economies of scope in the environmental management mechanisms. We found no evidence of an association between the average age of plant technology and superior environmental performance. The inverse relationship found between the export-to-sales ratio and environmental behaviour was not expected. While it is clear from previous studies that exporting firms fare better than non-exporting ones for manufacturing as a whole, there is no evidence of a statistically significant association (positive or negative) between the intensity of export activity and environmental performance. This result should be interpreted with caution as indicated above.

5 It must be stressed that the specific characteristics of the industry under study imply that our firms are in no way representative of the average firm in the chemical industry nor in Mexican manufacturing. Much research is needed in the case of small and medium-sized firms. However, as mentioned before, our results confirm those of previous studies, based on larger samples, by Domínguez (1999) and Dasgupta *et al.* (1997), in various aspects. First, we found that regulatory pressure is an important factor behind firms' environmental behaviour, and that communities can press in the same direction. Second, the role of formal systems of environmental

management, such as ISO 14000-type procedures mentioned by Dasgupta *et al.* (1997), was found to be associated with environmental performance, confirming the author's opinion that, although following the ISO 14000 sequence would not, of course, guarantee any improvement in environmental performance, it is likely that plants which complete these steps are better informed, organised and motivated to solve environmental problems. Last, we also found no association between technology vintages and environmental performance. In contrast, we did find positive results with multinational ownership.

Notes

1 The research for this paper was undertaken as part of a project funded by the Economic and Social Research Council, Global Environmental Change Programme on *Pollution Trade and Investment: Case Studies of Mexico and Malaysia* (Award No. L320253248).
2 See Appendix 1 for details of the firms interviewed.
3 National Association of Chemical Industry (ANIQ) Year Book, various years.
4 The main firms are: Celanese Mexicana, AKRA (Nylón de México and Fibras Químicas), Celulosa y Derivados (Cydsa), Fibras Sintéticas, Industrias Polifil, Finacril, Kimex and Impetmex.
5 Annual Industrial Survey, National Institute of Statistics and Geography (INEGI).
6 The Mexico City metropolitan area contributes 30 per cent of domestic residual water discharges, at a rate of 40,000 litres per second.
7 This complaint has been most frequent among small and mid-sized firms (see Domínguez, 1999).
8 Asociación Nacional de la Industria Química, http://www.aniq.org.mx.
9 Grupo Industrial Alfa (1997).
10 *Expansión* (1997).
11 One of the three absent firms did not agree to be interviewed and the other was discarded owing to its very low market share. Both firms are Mexican owned. Interestingly, the firm that refused to be interviewed participated in the 'Clean Industry' programme and has been certified.

Bibliography

Asociación de la Industria Química (ANIQ) (various years) *Anuario estadístico.*

Brown, F. and Domínguez, L.(1998) *Productividad: desafío de la industria mexicana.* México: Ed. JUS-UNAM.

Dasgupta, S., H. Hettige, and D. Wheeler (1997) *What Improves Environmental Performance? Evidence from Mexican Industry.* Washington, DC: World Bank, Development Research Group. Working Paper Series #1877.

Departamento de Distrito Federal (1992) *Balance Ambiental de la Industria en la Zona Metropolitana de la Ciudad de México*, Mexico, DF (mimeo).

Domínguez, L. (1995a) *Reconversión ambiental y Medio Ambiente: el caso de Dupont.* Seminario Instrumentos Económicos para un Comportamiento Empresarial Favorable al Medio Ambiente. México: El Colegio de México (mimeo).

Domínguez, L. (1995b) *Reconversión hacia Tecnologías Limpias: el caso de Cydsa.* Seminario Instrumentos Económicos para un Comportamiento Empresarial Favorable al Medio Ambiente. México: El Colegio de México (mimeo).

Domínguez, L.(1999) 'Comportamiento empresarial favorable al medio ambiente: el caso de la industria manufacturera de ZMCM', in A. Mercado (ed.) *Instrumentos económicos para un comportamiento empresarial favorable al ambiente en México.* México: El Colegio de México-Fondo de Cultura Económica.

Expansión (1997) Mexico, DF: August.

González Palomares, L. (1995) 'Análisis de la contaminación del agua en el caso de México' in J. Quiroz (ed.) *Análisis de la contaminación de aguas en América Latina.* Santiago de Chile: CENDE 1.

Grupo Industrial Alfa (1997) *Report to Shareholders.* Mexico.

Lexington Group, The (1996) *Industrial Environmental Management in Mexico: Report on a Survey.* Mexico (mimeo submitted to World Bank).

Máttar, J. (1994) 'La competitividad de la industria química', in F. Clavijo and J. Casar (eds) *La industria mexicana en el mercado mundial; elementos para una política industrial, LECTURAS*, Vol.II, No. 80. México: Fondo de Cultura Económica.

Mercado, A.(1995) *Competitividad y Ambiente: El caso de una maquiladora de componentes electrónicos.* Seminario Instrumentos Económicos para un Comportamiento Empresarial Favorable al Medio Ambiente. México: El Colegio de México (mimeo).

National Institute of Statistics and Geography (1998) *National Accounts.* Aguascalientes: AGS.

Quintana R.L, (1997) 'La industria mexicana de fibras químicas textiles', *Comercio Exterior*, No. 4.

Urquidi, V. L. (1972) 'Incentivos contra la contaminación', *La Gaceta*, No.17. México: Fondo de Cultura Económica.

Urquidi, V. L. (1994) *The Use of Economic Incentives for Clean Technology in Developing Countries*, paper presented at Third International Conference of the International Association for Clean Technology, Vienna, 6–8 April.

Urquidi, V. L. (1999) 'Instrumentos económicos para la política ambiental: Estructura industrial y comportamiento empresarial en los países en vías de desarrollo, con referencia a México', in A. Mercado (ed.) *Instrumentos económicos para un comportamiento empresarial favorable al ambiente en México*, México: El Colegio de México-Fondo de Cultura Económica.

Appendix 1: Characteristics of the sample

Of the nine firms[11] (thirteen plants) in the chemical fibres industry existing today, we interviewed six firms that have ten plants, contributing more than 90 per cent of the industry's total sales. In all cases their individual sales surpassed one billion pesos annually during 1998. Yet there are significant differences among them. Among the large firms in the sample, two are part of industrial conglomerates and another is a large corporation with several plants in the chemical industry. These three firms account for almost 85 per cent of domestic sales. There is foreign capital equity in two firms. The three remaining firms have only one plant each.

Half of the firms interviewed were founded in the initial phase of import substitution, i.e. between 1950 and 1965, and the other three began activities in the 1970s and 1980s. All firms have had important export activity since trade liberalisation began, which has grown since 1994, in part due to the collapse of the domestic market. Three plants export more than 50 per cent of their production (of four plants visited), and between 25 and 50 per cent of the production in the other two firms, which includes three plants exporting between 25 and 50 per cent of their production and one plant exporting between 5 and 25 per cent of its production. Export markets are basically North America, Latin America and Europe. Summarising, the sample covers firms of larger-than-average size in the chemical industry and with a considerable volume of exports.

Appendix 2: Methodology for measuring firm environmental performance

The evaluation of a firm's environmental performance was based on the questionnaires which were completed by the firms interviewed. The questionnaire was divided into four major sections dealing with environmental policies, environmental management, environmental programmes and activities, and environmental impacts. A distinction was also made between plant level and corporate management and programmes.

A scoring system was devised to evaluate the firm's performance under six different headings – policy; corporate management; plant management; corporate programmes; plant programmes; environmental impact. These scores were then aggregated to give overall environmental performance indicators at the plant level and for the company as a whole.

The scoring system was devised to reflect the extent to which a firm's responses to the questions were positive in terms of environmental performance. Thus, in the case of environmental policy a firm would score a maximum of 5 points if it had a written environmental policy; a commitment to environmental standards over and above those required by local regulation; had signed a national or international declaration on environmental principles; considered environmental matters to be a high priority; and had written procedures for dealing with environmental emergencies.

For environmental management, evaluation depended on the existence of a formal environmental management system, the existence of an exclusive management post with responsibility for environmental matters; an environmental management manual; indicators of environmental performance; frequency of monitoring; environmental incentives for staff; publication of information on environmental performance; procedures for dissemination of information on environmental matters within the firm.

Under firm-level programmes and activities, indicators used related the extent of such programmes; the existence of specific quantitative targets; training programmes; voluntary environmental audits; programmes with the local community.

In addition, scores were derived at the plant level which related to environmental management, reflecting the extent to which the firm had indicators of energy and water use, and of emissions to air and water. An evaluation was also made of the changing impact which the firm has on the environment through reductions in emissions and resource utilisation over time.

Adding together the scores obtained under the different headings gave overall evaluations of environmental performance, both at the corporate and plant level. In order to standardise the scores, which are based on a different total for each aspect, they were all expressed as percentages in the analysis.

10

ENVIRONMENTAL ASSESSMENT OF THE MEXICAN STEEL INDUSTRY

Alfonso Mercado[1]

Introduction

How to promote environmental concern on the part of firms has become an increasingly important issue. A prominent theme in recent discussions in Mexico has been the lack of economic instruments that motivate entrepreneurial behaviour favouring the environment (INE, 1996; Mercado, 1999). Another topic that has been discussed (and questioned) is the environmental behaviour of small firms, which often fail to comply with norms and environmental protection standards (Mercado and Fernández, 1998). In this debate, there is a consensus regarding the need to study the links between the development of a national economy that is increasingly geared towards other countries (especially the USA and Canada) and the problems of industrial pollution. Similarly, studies have been undertaken to assess the impacts of public environmental policies, and their mechanisms in the form of norms, other regulations and economic instruments. (Canela *et al.*, 1998; Mercado, 1999).

Among the manufacturing and service sectors that have received special attention in Mexico as a result of their environmental impact are the chemical, iron and steel, and hospital services industries. In Mexico, the basic iron and steel industry is the second most important manufacturing sector in terms of energy consumption, and is one of the major producers of industrial pollutants. Moreover, it is one of the sectors whose production and export growth has occurred at a considerably faster rate than the average for Mexican manufacturing as a whole.

In this chapter, we study to what extent Mexican environmental legislation and its enforcement have tended to mitigate the environmental damage associated with the industrial development of the steel industry. We also explore the influence of external linkages (through trade and foreign investment) on the structural problems of industrial pollution, as contrasted with the influence of the local framework of norms and laws. In other words, this research is geared to answering the following essential question: how do the economic conditions associated with increasing liberalisation of trade and investment flows in Mexico affect the environmental decisions made by steel companies, and how does the framework of norms and laws also affect them? This study also analyses the type of firms (by size, source of capital, external market orientation) that exhibit the best environmental management, more actively incorporate technologies that pollute less, make better decisions through programmes geared to curbing pollution, and have better environmental performance.

This chapter is divided into seven sections. In the first, we discuss the analytical framework of the study. In the second, we explain the research methodology. The third section analyses the technological options of that industry and their possible environmental impact. The fourth section is devoted to studying economic trends in the Mexican steel industry. Sections five and six include the main empirical findings of this study, consisting of an environmental assessment of the steel plants operating in Mexico from both a static and a dynamic standpoint. The final section presents the conclusions of this study, with an emphasis on the contextual factors that have a bearing on environmental assessments.

Analytical framework

When analysing the environmental behaviour of firms in a business world tending towards globalisation and integration, it is useful to distinguish three separate effects (see Grossman and Krueger, 1992; Chapter 2, this volume). First, there is a *scale effect*, which is derived from the impact on the level of economic activity, especially the level of industrial output. Second, there is a *composition effect*, which reflects changes in the shares of the different industries in total industrial production. Last, there is a *technological or process effect*, which is related to changes in the intensity of each industry's level of pollution (i.e. the average volume of pollution per tonne or litre – or some other unit of physical measurement – of output).

These effects can be produced due to the establishment of norms and other command and control instruments, as well as economic instruments such as changes in utility prices, anti-pollution taxes and preferential credits. They may also be generated by greater receptivity to foreign investment flows and imports, and may even be promoted through external demand in the form of exports.

Up to now, evidence has been limited and often contradictory. It is hard to reach conclusions regarding the scale, composition and technological effects in Mexican industries, given the various macro- and micro-economic contextual changes that have occurred. Without proper research in Mexico, it is not possible to have a clear notion of this or of other critical aspects such as compliance with environmental laws and norms related to the industry, the impact of new economic instruments with environmental objectives and the effectiveness of public policies and pro-environmental initiatives undertaken by firms. In addition, little is known about the co-ordination problems among the different Mexican authorities involved, at the municipal, state and federal levels, in the design, application, monitoring and execution of sanctions set by law.

This study does not propose to provide answers for all these issues, but rather to gain further knowledge regarding the experience of the steel industry, which is very important in Mexico, from both an economic and environmental perspective. More specifically, we intend to study whether plants which tend to adopt cleaner technologies and are prone to complying above and beyond what is required of them by Mexican norms have the following characteristics: (a) they are subsidiaries of transnational companies; (b) they belong to large Mexican companies; (c) they are large plants; (d) they are major exporters. The initial hypothesis is that plants with these characteristics generally exhibit better environmental management and performance owing to their competitive, organisational, financial and technological advantages. Moreover, as a result of strategic and market factors, we feel that they are more sensitive to the influence of contextual factors such as the liberalisation of the economy, changes in public environmental policies (with strict requirements to comply with the norms and the introduction of economic instruments) and pressure from society and from other countries.

Methodology

As stated above, this chapter will offer two types of analysis. One is aggregate and deals with economic and technological trends in the Mexican steel industry and their environmental impact. The second type of analysis is microeconomic and focuses on an environmental assessment of steel plants. This assessment, in turn, adopts two approaches: one that is static, and the other, dynamic.

The study of the environmental impact of the Mexican steel industry is based on the construction of indicators and a review of technical reports. The indicators are energy intensity (energy consumption per tonne of output) and intensity of industrial wastewater discharges (cubic metres of industrial wastewater discharges per tonne of output).

Assessment at the microeconomic level is based on a survey which collected data on the plants through direct interviews, using a questionnaire. For this micro-scale assessment, a methodology of assigning points to the answers given by the firms was adopted. Its logic is simple: to assign the highest score to the firms which have the most developed environmental policies and management systems, the most comprehensive environmental programmes, and the best environmental performance at the firm and plant levels. The scoring system was based on a checklist incorporated in the questionnaire. The answers to this checklist are weighted, with the objective of clearly distinguishing between a set of plants with a good environmental performance and another set of poorly performing plants.

The evaluations of firm performance at a point in time were grouped into two sets of items: the corporate level and the plant level. The dynamic evaluation of firm behaviour involved assessing trends in environmental policies and actions over time. These elements were grouped into three issues: (a) productive changes for environmental purposes, (b) the evolution of the plant's environmental performance, and (c) improvements in environmental management (see Appendix 1 for more details on the scoring system applied). A high score in the static evaluation indicates a good level of current performance, while those with a high dynamic score may be considered as successful in their environmental progress in terms of policy, management, actions and performance. On the other hand, those plants with a low dynamic score may be considered as lagging behind in their environmental performance and behaviour over time.

Process technology and the environment

The steel industry is one of the most polluting manufacturing activities in relation to output. World Bank estimates of pollution intensity for 80 International Standard Industrial Classification (ISIC) four-digit industries, based on US data, puts the iron and steel industry in the top ten for most of the pollutants listed. For atmospheric pollution, it is the fifth most polluting industry in terms of sulphur dioxide emissions per million dollars of output, seventh for nitrogen dioxide, third for carbon monoxide, ninth for particulates and third for fine particulates (PM10). In water it is the most polluting industry for total suspended solids, while, for toxic pollution, iron and steel ranks between fourth and eleventh depending on the weightings used (Hettige *et al.*, 1995). The environmental challenge for the steel industry is, therefore, enormous, making it imperative to cut back on its pollution indices.

The steel production process and its environmental impact have already been described in Chapter 5 of this volume, and it is not necessary to repeat that discussion here. This section will concentrate rather on some of the specific characteristics of the Mexican steel industry, which have environmental implications. As was noted in Chapter 5 different stages of the production process have different levels of pollution associated with them, while different production technologies also have different implications in terms of the types and levels of pollution to which they give rise. Primary steel making, which involves converting iron ore into steel, has the most polluting processes, particularly in integrated plants, which include sinter and coke production. Production of finished products is often sub-divided into rolling of long and flat products, and a specialised sector that produces tubes and pipes.

Unlike Brazil, a significant proportion of steel production in Mexico is based on direct reduction of iron (DRI) using natural gas rather than the more conventional blast furnace using coke. DRI processes were first introduced on an industrial scale in the late 1950s, and one of the leading processes in use worldwide today was developed by Hylsa in Mexico. The process uses natural gas as the feed material to produce the reducing gas that is used to reduce iron ore to sponge iron in a shaft furnace. DRI avoids two of the most polluting processes normally involved in integrated steel making – coke production and sintering. It is, therefore, considerably less polluting than the blast-furnace route. It also requires less energy overall per ton of iron produced. In the late 1980s about a third of Mexican production was based on direct reduction, and by 1997 this proportion had risen to a half (see Table 10.1).

In the steel-making process, open-hearth operations were phased out in Mexico in 1991. During the 1990s, the trend had been for the proportion of output accounted for by electric arc furnaces (EAFs) to increase from less than half of total steel production in the late 1980s to almost two-thirds of the total by the late 1990s (Table 10.1). Although the EAF is not necessarily less polluting than the basic oxygen furnace, an EAF works produces considerably less pollution than the integrated route. This reflects the fact that such mini-mills use scrap and/or sponge iron produced by direct reduction, whereas an integrated works will include a sinter plant and coke oven. Thus, the shift from the integrated route to EAF should help reduce the average level of pollution intensity in the industry.

The third major change in the industry has been the growth of continuous casting, which has increased its share of total semi-finished products produced in Mexico from just over half in the late 1980s to 85 per cent by the late 1990s

Table 10.1 Shares of DRI, EAF and continuous casting in Mexican iron and steel production, 1988–97

	DRI (%)	*EAF* (%)	*CC* (%)
1988	31.4	45.8	55.9
1989	40.1	51.8	57.6
1990	40.8	51.4	63.2
1991	44.9	57.5	64.4
1992	40.5	55.7	64.4
1993	44.4	59.2	76.3
1994	47.9	62.6	74.9
1995	47.2	62.6	76.9
1996	47.3	64.1	84.9
1997	49.9	65.2	84.9

Source: IISI (1999): Tables A-1, A-7, B-4 and B-8.

(Table 10.1). Continuous casting is less polluting than ingot casting and also requires less energy per tonne of steel.

Thus, the process changes which have taken place in the Mexican steel industry since the late 1980s have been in the direction of less polluting and less energy-intensive processes and this should have reduced the environmental impact of the industry per tonne of steel produced. Unfortunately, there is limited direct information on the level of pollution generated by steel production in Mexico. However, there are data on the use of energy by the industry, which can give some idea of trends in energy intensity and may be a proxy for pollution intensity as well.

The steel industry is one of the most energy-intensive manufacturing branches in Mexico. On average, the energy intensity of the industry is more than four times the manufacturing average, and its share of total energy consumption by the manufacturing sector increased from 15.6 per cent in 1992 to 19.3 per cent in 1997 (see Table 10.2). Nevertheless, there has been a significant reduction in energy intensity in the steel industry since 1995.

The Mexican steel industry is among those that most pollute the water and it is also among those that have done the most to carry out projects on environmental hazards caused by industrial activity. Tables 10.3 and 10.4 indicate that, in 1993, this industry was in third place as regards the production of wastewater (after the sugar and chemical industries), producing on average the equivalent of 75 per cent of the discharges attributable to the chemical industry and 17 per cent of those attributable to the sugar industry. In the Balsas River–Infiernillo watershed, the base metal industry has produced more BOD than the chemical, beverage and petrochemical industries.

Economic trends

In the past ten years, major changes have taken place in the Mexican steel industry. It went through both a privatisation process and increasing

Table 10.2 Energy intensity in the Mexican steel industry compared with the manufacturing sector

	Manufacturing[a]	*Iron & steel*[a]	*Share of Iron & Steel*[b]
1992	5.04	27.51	15.6
1993	5.18	26.38	15.5
1994	5.26	27.00	16.3
1995	5.77	27.51	17.7
1996	5.32	25.62	18.5
1997	4.86	23.97	19.3

Source: INEGI (1999a, b).

Notes

a Petajoules of energy per thousand pesos of GDP at 1993 prices.

b Share of the iron and steel industry in total energy consumption by the manufacturing sector.

Table 10.3 Production of wastewater and BOD in the Balsas River–Infiernillo watershed in 1993

	No. of plants	*Volume*		*BOD*	
		m³/day	*%*	*kg/day*	*%*
Total	9	1,039,067	100.0	132,254	100.0
Base metals	4	400,426	38.5	73,747	55.8
Chemicals	1	587,720	56.6	57,430	43.4
Beverages	1	281	0.0	99	0.1
Petrochemicals	1	80	0.0	2	0.0
Others	2	50,760	4.9	977	0.7

Source: Comisión Nacional del Agua. Internal document SEMARNAP (1997).

liberalisation of trade and investment. Steel production grew at higher annual rates than most of Mexican manufacturing production, reaching two-digit annual growth rates (see Table 10.5).

In 1996 the industry accounted for 4.3 per cent of manufacturing gross domestic product (GDP) and employed 1 per cent of workers in manufacturing. Between 1988 and 1997, Mexican steel production grew by 78 per cent, despite a contraction in 1991. Contrary to the overall trend observed in production, employment in the sector decreased during this ten-year period by 58 per cent. This led to a significant increase in labour productivity, especially between 1989 and 1990 and between 1991 and 1992 (see Table 10.6). In the first of these two periods, there was a major rise in production, accompanied by a decrease in employment; in contrast, during the second period production rose only slightly, but the contraction in employment in this sector was the greatest recorded during the period (Table 10.6).

Total factor productivity (TFP) has been growing too. Using estimates based on the annual industrial survey done by the Mexican National Institute of Statistics, Geography and Informatics (INEGI), Brown and Guzmán (1998) found that, from 1984 to 1994, the TFP of the basic iron and steel industry tended to increase. Despite some fluctuations between 1984 and 1988, the TFP of this sector continued to rise, each year, from 1988 to

Table 10.4 Mexico: intensity of industrial wastewater discharges, 1993

Sector	*Discharges (%)*	*GDP (%)*	*Intensity (discharges/GDP)*
Sugar	39.4	1.1	35.8
Chemicals	18.6	11.2	1.7
Base metals	6.8	3.7	1.8

Source: Comisión Nacional del Agua. Internal document SEMARNAP (1997).

Table 10.5 Annual growth rate[a] of the iron and steel industry's GDP, 1992–7

Sector	*1992–3*	*1993–4*	*1994–5*	*1995–6*	*1996–7*
Manufacturing industry	−0.67	4.07	−4.94	10.83	9.96
Iron and steel	5.40	8.99	11.02	14.91	11.70

Source: INEGI (1999).
Note
a As a percentage – based on value of the GDP in millions of pesos at 1993 constant prices.

1994. This trend was most marked in the segment of primary steel making, with an average annual growth rate of 14.2 per cent. Productivity growth was less rapid, but still significant in rolling (with an average annual growth rate of 8.3 per cent) and in the manufacture of steel tubes (9.4 per cent). The highest TFP growth rate of the first two segments mentioned occurred in giant firms, while the fastest growth of the third segment took place in large (but not giant) firms and, quite close to these, in small firms (Brown and Guzmán, 1998: 843).

During this period, the productivity gap between large firms and medium-sized and small ones has been increasing, particularly in primary steel making and rolling. The exception seems to be the segment of steel tubes, in which productivity convergence between small to medium-sized plants and large plants is estimated. Firms' reaction to market pressures, resulting from the increasing liberalisation, have induced these trends. On the one hand, the

Table 10.6 Output, employment and labour productivity in the Mexican steel industry, 1988–97 (base 1993 – 100)

Year	*GDP (index)*	*Employment (number of workers)*	*Labour productivity*[a] *(index)*
1988	87.3	81,731	38.4
1989	89.4	74,732	42.9
1990	98.0	63,909	55.1
1991	93.4	55,546	60.4
1992	94.9	45,265	75.3
1993	100.0	35,921	100.0
1994	109.0	34,103	114.8
1995	121.0	32,799	132.5
1996	139.1	34,128	146.3
1997[b]	155.3	35,528	157.0

Source: INEGI (1997, 1998, 1999a).
Notes
a Calculated on the basis of the following formula: $[(GDPK/PE)_i/(GDPK/PE)_b]\times 100$, where GDPK is the GDP at constant prices, PE is the number of persons employed, *i* is the year being analysed, and *b* is the base year, 1993.
b Preliminary figures.

plants dominating the market were restructured and modernised. Exports of iron and steel products tended to rise, as did the major indicators of efficiency, productivity and quality. Therefore, the plants' value added grew considerably, while the employment of workers decreased (Table 10.6). The financial costs involved in these investment efforts are reflected in increasingly poor indicators for the leverage ratio and liquidity. On the other hand, medium-sized and small firms lagged behind in terms of modernisation and efficiency. They tended to service domestic niche markets, and their exports have been minimal. Some of these plants gear their production to local markets in the centre and south of the country.

The steel industry can be divided into three different sub-sectors, primary steel making, including the casting of semi-finished products; hot and cold rolling of flat and long products; and production of steel tubes. In Mexico the first two each accounts for over 40 per cent of the steel industry's GDP, while tubes account for between 10 and 15 per cent. The structure of steel production in Mexico changed gradually in the 1990s. The share of primary steel making has increased from 40 to 45 per cent, while that of rolling has moved in the opposite direction (see Table 10.7). The most striking change in the industry, however, has been the growth of exports, particularly after 1994 when exports increased more than 150 per cent in both dollar and volume terms.

The growth of steel exports was concentrated in semi-finished steel products during the first half of the 1990s, but the spectacular expansion in 1995 was mainly of finished products. It appears that the NAFTA induced a strong structural change in the exported product mix. Before NAFTA, in 1992, the main steel goods exported were semi-finished products. Since 1995, the major steel articles exported have been finished products (in 1997 they accounted for 61 per cent of the volume of exports) while the export share of primary

Table 10.7: Shares of major sub-sectors in steel industry GDP, 1988-97

	Semi-finished products (%)	*Finished products (%)*	*Steel tubes (%)*
1988	41.6	43.7	14.7
1989	41.9	43.8	14.3
1990	41.9	44.3	13.8
1991	40.3	44.1	15.5
1992	41.3	45.7	13.0
1993	41.1	46.6	12.3
1994	41.7	45.6	12.6
1995	43.1	45.5	11.5
1996	45.0	41.8	13.2
1997	43.5	41.2	15.3

Source: INEGI (1999).

products fell to 39 per cent of the physical volume of the sector's exports (see Table 10.8).

Looking at imports, the trend has been in the opposite direction to that of exports in terms of both composition and volume. Between 1992 and 1997, the import of steel declined, and its composition remained predominantly of finished and final consumption products. As a consequence of these trends the trade balance of iron and steel products is positive, especially in the case of primary products.

The change in the composition of exports has not had a major impact on the structure of production in Mexico. If anything, as Table 10.7 indicates, the share of primary steel making in industry value added has increased, despite the fall in the share of semi-finished products in the industry's exports. This, together with the very low level of imports of semi-finished steel products to Mexico, indicates that the growth of exports of finished steel products is based on integrated domestic steel production. Thus, exports of finished steel products have led to the growth of all steel-making activities in Mexico, including the primary processes, which as was seen in the previous section are the most polluting.

In recent years, the Mexican steel industry has been facing major financial and profitability problems. For instance, the ratio between net sales and total assets, dropped from 1990 to 1995, reaching its lowest level, 0.29, in 1994.

Table 10.8 Exports and imports of iron and steel products in Mexico, 1992–7

Year	*Total Value (US$ million)*	*Total Volume (tonnes)*	*Semi-finished iron and steel products (tonnes)*	*%*	*Finished steel products (tonnes)*	*%*
Exports						
1992	1042	1,730,879	930,569	53.76	799,310	46.18
1993	895	2,155,732	1,356,468	62.92	799,264	37.08
1994	1088	2,371,101	1,494,245	63.02	876,856	36.98
1995	2717	6,276,668	2,143,003	34.14	4,133,665	65.86
1996	2713	5,757,862	2,105,616	36.57	3,652,246	63.43
1997[a]	3210	5,995,228	2,343,013	39.08	3,652,215	60.92
Imports						
1992	2237	3,037,987	194,897	6.42	2,843,090	93.58
1993	1614	1,762,632	133,248	7.56	1,629,384	92.44
1994	2733	3,102,674	153,754	4.96	2,948,920	95.04
1995	950	971,727	65,504	6.74	906,223	93.26
1996	1228	1,111,228	86,726	7.80	1,024,502	92.20
1997[a]	1547	1,658,505	224,328	13.53	1,433,977	86.46

Source: INEGI (1997, 1998, 1999a).

Note

a Preliminary figures.

The leverage ratio also worsened; for example, the ratio between total liabilities and total assets increased by 70 per cent (from 0.39 to 0.66) in the five years under consideration, while the ratio between net sales and total liabilities fell to less than half (from 1.2 to 0.51). Liquidity, quantified by the ratio between current assets and current liabilities, dropped sharply from 1.31 in 1990 to 0.56 in 1995. The problem of this sharp fall in liquidity was heightened as regards stock credits, going from a ratio of 21.88 in 1990 between current assets and short-term stock credits to 1.88 in 1995, while the greatest contractions occurred in 1992 and 1995. The rate of return (net profit for every peso of net sales) was very unstable in 1990–3, and negative in 1994 and 1995 (INEGI, 1997, 1998, 1999a).

Environmental assessment of 12 Mexican plants: static approach

General trends indicate that the Mexican steel industry is expanding rapidly, modifying its product mix, increasing productivity and reducing energy intensity, in a context of increasing liberalisation through the NAFTA. It is also known that there are important structural differences in these trends, between segments of the steel sector and types of plant. A deeper analysis is now necessary, particularly in relation to the environment. The objective of this section is to assess the environmental behaviour of twelve selected steel plants in Mexico, in the economic and environmental context mentioned above, and in the socio-political context of public (and foreign) pressures to reduce pollution intensity. First, an explanatory note of the sample interview is presented, and then the main findings of the static study are discussed.

The sample and selection criteria

Our interest in studying the Mexican steel industry focuses on its most polluting production processes, i.e. steel making and rolling. These are the processes for making basic iron and steel products. Aside from this central interest, we are also concerned with carrying out a comparative analysis of other downstream processes for manufacturing finished steel products such as drawing-wire rods and rods, and cutting and shaping structural pipes and sections, as well as intermediate and final chemical treatment by galvanising sheet metal. This order of priorities made it imperative to be stricter in selecting the sample of steel and rolling mills (regardless of their degree of integration) than in the case of other non-integrated plants such as foundries and terminal producers of tubes, wire rods, rods, structural sections and other finished products. On the other hand, the author felt that it was likely that firms would refuse to answer the survey questionnaire for various reasons, among which the most important have to do with firms' perceptions concerning issues of pollution and environmental actions in Mexico. First, the topic involves confidential information that can even be compromising for the company. Second, the topic is not one of the urgent priorities that most firms have to attend to in the current situation of the industry, which is undergoing a recessionary crisis. Under these circumstances, the sample cannot be random. Thus, we decided to select two groups of firms: (1) the major steel companies with steel making and rolling, and (2) a variety of companies that effect any of the following processes: steel making, rolling and product finishing (for further details on the sampling frame and the final sample, see Appendix 2).

Among the different firms surveyed, the majority are relatively large (in comparison with other industrial sectors in Mexico), integrated firms with Mexican capital and technological levels that are not very advanced. The sample is made up of nine integrated or semi-integrated firms with facilities for steel making and three non-integrated firms. In terms of size, the sample includes five large firms and seven medium-sized ones by steel industry standards. Technological levels vary, with four of the twelve firms reporting state-of-the-art technology. Four firms export over 50 per cent of their sales, six export between 5 and 50 per cent, and two less than 5 per cent. Nine of the firms are 100 per cent nationally owned, one has mixed (national and foreign) capital, and two have 100 per cent foreign capital (Table 10.9).

Static environmental assessment: research findings

The main pollution problems of this set of twelve steel plants are air pollution (in eleven of these plants), solid waste management (nine of these plants) and water contamination (six of these plants). These environmental problems have been faced by plant managers in a context of increasing volume of production, with different approaches and results, as will be seen below.

In the group of firms surveyed, the majority reported changes in scale of production and technology, but none had significant changes in the composition of output (see Table 10.10). This is consistent with the aggregate data presented earlier since the product mix has not been modified as much as the volume of output and technology (as indicated by the growth in labour productivity). However, at the same time, the earlier figures also indicate that the growth of output in iron and steel has been much greater than the reduction in energy intensity, so that the same is likely to have been true of pollution intensity as well. Thus, it can be expected that the dominant factor within the steel industry has been the scale effect (environmentally adverse) rather than the technological effect (environmentally favourable). Unfortunately, we cannot verify this perception, either in terms of the general trend or in terms of the extent of reductions in specific emissions.

As regards environmental measures taken, the average overall score is in a medium range, 6.1 points out of a theoretical maximum of 10, and the maximum ranking achieved by a firm was 9.0, as can be seen in Table

Table 10.9 Principal features of the sample

Features	*No. of firms*
A. Size	
Large (over 1000 workers)	5
Medium (less than 1000 workers)	7
B. Processes[a]	
Integrated or semi-integrated	9
Non-integrated	3
C. Technology[b]	
Level I	4
Levels II, III and IV	8
D. Export level	
High exports (over 50% of sales)	4
Medium exports (5–50%)	6
Low exports (less than 5%)	2
E. Source of capital	
Foreign	2
Mixed national–foreign	1
100% national	9
Total	12

Source: Firm survey.

Notes

a 'Integrated' or 'semi-integrated' processes include steel making, while 'non-integrated' processes do not.

b Technological levels: I. Comparable with the best existing level at present. II. Best existing level from 2 to 5 years ago. III. Best existing level from 5 to 10 years ago. IV. Best existing level from 10 to 20 years ago.

Table 10.10 Productive changes made by ten steel firms

Type of change	*No. of firms*
Scale	8
Composition	0
Technological	8

Source: Firm survey.

10.11. In this table, we present average scores and the maximum and minimum scores obtained by the different firms. A couple of findings are worth looking into. First, the great variance between the maximum and minimum scores: there are firms with the highest possible score (10.0) and

Table 10.11 Evaluation of the environmental protection by 12 steel firms, 1998

	Rating		
Concept	*Average*	*Maximum*	*Minimum*
A. Corporate level (average)	(6.3)		
Firm's Environmental Policy	6.9	10.0	2.5
Environmental programmes and activities	5.6	10.0	1.3
Environmental management	6.4	10.0	2.0
Environmental performance of the firm	6.3	10.0	2.5
B. Plant level (average)	(5.7)		
Environmental programmes and activities	5.2	10.0	2.5
Environmental performance of the plant	6.1	10.0	4.3
C. Overall (average)	(6.1)	9.0	3.6

Source: Firm survey.

others with scores as low as 1.3 to 4.3. That illustrates the great heterogeneity of environmental policies, actions and performance in the industry. Second, the large difference between the relatively low score for pro-environmental actions (5.2 for plants and 5.6 for firms) and the relatively high score for the environmental policies of the firm (6.9). Therefore, it appears that, at present, greater emphasis is placed on formulating environmental policies than on taking effective actions.

In view of the high degree of heterogeneity of environmental behaviour and performance, it is worthwhile considering what type of firms tend to be most proactive environmentally under present conditions. Despite the limitations of the sample, and without attempting to arrive at definitive conclusions, the data in Table 10.12 suggest that large steel firms tend to be more environmentally conscious than medium-sized ones; firms with a greater degree of high technology do more than those with technology from earlier years; export-oriented plants (high and medium exporters) do more than those oriented to the domestic market, and those that have integrated or semi-integrated processes (with a steel mill) do more than non-integrated ones. The grouping of firms by source of capital and by degree of exports[2] does not allow for a differentiated appraisal of their environmental behaviour, suggesting that these variables are not very relevant. This implies that foreign investment does not seem to have a direct bearing on the environmental behaviour of firms in the steel industry. The better environmental scores achieved by large plants, plants with high technology, export-oriented plants, and ones with integrated or semi-integrated

Table 10.12 Environmental evaluation of group of 12 steel firms by characteristics, 1998

		Corporation level					Plant level			
Characteristics of firm	*No. of firms*	*Environment policy*	*Actions with programmes*	*Management*	*Environment performance*	*Sum*[a]	*Actions with programmes*	*Environ. performance*	*Sum*[a]	*Overall*[a]
Size[b]										
Large	5	8.6	7.5	8.3	6.0	7.6	6.0	7.1	6.6	7.3
Medium	7	5.7	4.2	5.1	6.4	5.4	4.6	5.3	5.0	5.2
Process[c]										
Integrated or semi-integrated	9	7.6	6.4	7.0	6.1	6.8	5.3	6.3	5.8	6.4
Non-integrated	3	4.8	3.2	4.5	6.7	4.8	5.0	5.2	5.1	4.9
Technology[d]										
Level I	4	7.0	8.5	7.7	5.6	7.2	6.9	7.9	7.4	7.3
Levels II, III and IV	8	6.9	4.1	5.8	6.6	5.9	4.4	5.2	4.8	5.5
Exports[e]										
High	4	7.8	5.9	6.2	5.6	6.4	4.3	6.0	5.2	6.0
Medium	6	6.8	6.6	7.0	6.7	6.8	5.8	6.4	6.1	6.5
Low	2	5.6	1.9	5.2	6.3	4.8	5.0	5.0	5.0	4.8
Source of capital										
Foreign[f]	3	7.7	7.4	6.7	5.8	6.9	5.8	5.2	5.5	6.4
100% national (private)	9	6.7	5.0	6.3	6.4	6.1	5.0	6.3	5.7	6.0
Total	12	6.9	5.6	6.4	6.3	6.3	5.2	6.1	5.7	6.1

Source: Firm survey.

a Arithmetic average of the corresponding items in each row.

b Large firms, with over 1000 employees, and medium-sized ones, with 100–1000 employees.

c An integrated process includes the primary process (casting, steel making) up to the final manufacturing process. A non-integrated process does not include a primary process. It may be a secondary process (rolling) and/or a final one (the manufacture of tubes, sections, etc.).

d Technology levels:
I. comparable with the best existing level at present; II. best existing level from 2 to 5 years ago;
III. best existing level from 5 to 10 years ago; IV. best existing level from 10 to 20 years ago;

e High exports refer to over 50% of total sales, medium exports to 5–50% of sales, and low exports to less than 5% of sales.

f Two plants belong to 100% foreign firms, and one to a mixed national–foreign firm.

processes reveals that four variables play a central role in a firm's behaviour regarding industrial pollution:

1 *Size* – possibly associated with important economies of scale and scope in pollution control, and with advantages in financial, human and organisational resources.
2 *Foreign markets* – requiring standards of quality and environmental protection.
3 *Technology* – possibly associated with a favourable technological culture and access to restricted technical information concerning pollution control.
4 *Type of process* – associated with greater environmental hazards and greater intensity of pollution by primary processes, such as steel making, which require more attention to be paid to the environment.

We asked the firms about the major factors that have motivated them to take environmental measures, and the main response was governmental regulations and, in second place, demands imposed by foreign customers. The firms were also asked about the major obstacles they have confronted in terms of environmental measures. Most of the responses coincide in clearly pointing out two obstacles that are external to the firm, i.e. the high cost of anti-pollution equipment and efficient and 'environmentally friendly' equipment, and the lack of pro-environmental incentives. In terms of obstacles that are internal to the firm, the answers were not so consistent, but it is clear that their greatest internal problems are the existence of other more pressing priorities for the firm and the lack of technical and environmental knowledge and information (see Table 10.13).

The sample was separated into two groups: on the one hand, firms with a better environmental performance (five) and, on the other, those with a poor score (seven). Comparing the two groups revealed differences in terms of their major motivations and the main obstacles to undertaking environmental measures. After that review, we found two interesting motivations. First, although government regulation is the main motive for both groups of firms (Table 10.14), foreign market pressure is a more important factor amongst firms that perform well than for those who do badly. Second, government environmental regulation is a more important motive for poor performers (six out of seven) than for good performers (three out of five). The main external and internal obstacles of firms that are lagging behind in their environmental performance were:

- high equipment costs
- the lack of a supply of 'clean' or 'anti-pollution' technologies
- the need to attend to other more urgent priorities

Table 10.13 Perception of contextual factors affecting firms' environmental performance

Factors	*Frequency*
A. Factors favouring it	
Governmental regulations	9
Demands of foreign customers	3
Own policy by conviction	3
Public image or social pressures	3
B. Internal obstacles	
Other urgent priorities	4
Lack of environmental knowledge	4
Lack of technical information	3
Insufficient funding	1
Lack of trained personnel	1
No internal obstacles	3
C. Obstacles external to plant	
High equipment costs	8
Lack of incentives	7
Lack of technology	1
No external obstacles	2

Source: Firm survey.

Table 10.14 Main motives[a] to protect the environment by type of plant, according to their environmental evaluation (absolute frequency of answers)

Motive	*The 12 plants*	*The 5 best evaluated*	*The 7 poorest evaluated*
Environmental regulation	9	3	6
Influence from clients abroad	3	2	1
Public image	3	2	1
Own philosophy, sense of responsibility	3	1	2

Source: Firm survey.
Note
a Firms were allowed to identify more than one main motive.

- a lack of incentives and of technical and environmental information (see Table 10.15)

In short, the survey indicates that the large firms, oriented to foreign markets and with recent technology, are more successful in their environmental measures.[3] This is in a context in which government regulations and the demands of foreign markets seem to exert effective pressures on firms to reduce pollution, but where firms face cost problems or lack of technical–

Table 10.15 Main obstacles to environmental protection by type of plant, according to their environmental evaluation

Obstacles	*The 12 plants*	*The 5 best evaluated*	*The 7 poorest evaluated*
A. External			
High cost of equipment	8	2	6
Lack of incentives	7	2	5
B. Internal			
Other priorities which need urgent attention	4	0	4
Lack of technological information	3	1	2
Lack of environmental knowledge	4	2	2

Source: Firm survey.

environmental information, and other urgent problems that keep them from achieving better environmental performance.

Environmental evaluation of a group of twelve Mexican plants: a dynamic approach

This section presents a dynamic environmental assessment of twelve Mexican plants. The environmental evolution of this group of plants has been very disparate, and it has consisted mostly of 'end-of-pipe' solutions. Within the group, managers in eight plants stated that the main area of environmental progress had been the treatment of discharges and emissions. In five plants they said that there had been major progress in the implementation of environmental management systems. Five plants said that a major area of progress had been waste recycling. Only three plants mentioned better maintenance of equipment, and only two plants cited process innovation with environmental objectives.

Table 10.16 summarises the assessment of the evolution of environmental behaviour. First of all, two questions are interesting. On the one hand, there is a general lack of progress in the area of changes in production and technology for environmental purposes, environmental performance trends, and improvements in environmental management. On the other hand, the evolution of environmental behaviour is extremely heterogeneous, with high performance plants (reaching 10 points) and poor performance plants (between 0 to 2.5 points). This dispersion is even higher than in the previous static evaluation.

Looking at changes in production and technology for environmental purposes, the least important changes according to managers have been in product mix and production technology. The important changes seem to be related to input substitution (using fewer and less highly polluting inputs)

Table 10.16 Assessment of the evolution of environmental behaviour

Type of change	*Average rating*	*Maximum rating*	*Minimum rating*
A. Changes with environmental objective	4.6	10.0	2.5
B. Evolution of environmental performance	5.0	10.0	1.7
C. improvements in the environmental management	5.8	10.0	0
Overall (average)[a]	4.9	9.1	1.8

Source: Firm survey.
Note
a The weighted average is as follows: the type of change A = 36.4%; B = 54.5%; C = 9.1%.

and, most of all, the adoption of new, anti-polluting equipment. On the other hand, from a dynamic perspective, there has been a relative lack of progress in terms of the *quality* of emissions from the plants, while the main progress has been made in reducing the *volume* of resources (energy, water) (see Table 10.17).

Table 10.18 provides useful information concerning the characteristics of those steel plants that are achieving the greatest environmental progress. The large plants with more advanced technology, integrated or semi-integrated and export-oriented (in terms of the volume of exports) and foreign owned are more successful in their environmental progress, according to this

Table 10.17 Progress in environmental behaviour

Type of changes	*Frequency (%)*
A. CHANGES WITH ENVIRONMENTAL OBJECTIVE	
Technical change in the equipment	75
Less polluting input substitution	50
Innovation in technological process of production	50
Changes in product mix	8
B. EVOLUTION OF ENVIRONMENTAL PERFORMANCE	
B.1 Reductions in resource use	
Lower fuel oil intensity	50
Lower electricity intensity	75
Lower water intensity	67
Lower volume of wastewater discharges	50
B.2 Improvements in quality of emissions	
Better quality of wastewater discharges	42
Better quality of emissions to the atmosphere	42

Source: Firm survey.

Table 10.18 Assessment of progress in environmental behaviour of a group of 12 steel firms by characteristics, 1998

Characteristics of firm	*No. of firms*	*Evolution of the environmental behaviour*			
		Changes with environmental objective	*Evolution of environmental performance*	*Improvements in environmental management*	*Overall*[a]
Size[b]					
Large	5	5.5	6.6	8.0	6.7
Medium	7	3.9	3.8	4.3	4.0
Process[c]					
Integrated or semi-integrated	9	4.7	8.2	6.7	6.5
Non-integrated	3	4.2	3.6	3.3	3.7
Technology[d]					
Level I	4	7.5	7.5	7.5	7.5
Levels II, III and IV	8	3.1	3.8	5.0	4.0
Export ratio					
High (50% +)	4	3.1	5.6	7.5	5.4
Medium (5–50%)	6	6.3	5.7	5.0	5.7
Low (0–5%)	2	2.5	3.8	5.0	3.8
Export volume					
Large volume[e]	5	5.5	6.6	8.0	6.7
Smaller volume	7	3.9	3.8	4.3	4.0
Source of capital					
Foreign[f]	3	5.0	3.9	10.0	6.3
100% national (private)	9	4.4	5.2	4.4	4.7
Total	12	4.5	5.9	5.5	5.3

Source: Firm survey.

Notes

a Arithmetic average of the corresponding items in each row.

b Large firms, with over 1000 employees, and medium-sized ones, with 100–1000 employees.

c An integrated process includes the primary process (casting, steel making) up to the final manufacturing process. A non-integrated process does not include a primary process. It may be a secondary process (rolling) and/or a final one (the manufacture of pipes, sections, etc.).

d Technology levels:

I. comparable with the best existing level at present;
II. best existing level from 2 to 5 years ago;
III. best existing level from 5 to 10 years ago;
IV. best existing level from 10 to 20 years ago.

e Large volume of exports refers to high and medium export ratios by large plants.

f Two plants belong to 100% foreign firms, and one to a mixed national–foreign firm.

assessment. It is particularly interesting to note that the plants with a low ratio of exports (exporting less than 5 per cent of production) are the group with the poorest environmental assessment. On the other hand, the plants with a high export ratio (exporting more than 50 per cent of production) are not the best performers, but rather those with intermediate ratios (exporting between 5 and 50 per cent of production). However, it is the firms with the largest absolute volume of exports which were the best performers. Furthermore, Table 10.18 reveals that large plants, with significant export volumes, are the best performers in terms of their progress in environmental management. Vertically integrated plants have the best assessment in relation to the evolution of environmental performance. The plants with a high technological level achieved the highest score for productive changes with environmental purposes, and also had a high assessment concerning both the evolution of environmental performance and improvements in environmental management. Table 10.18 also indicates that foreign ownership is associated with greater progress in environmental performance, suggesting that economic and technological linkages with foreign corporations may induce or facilitate such progress.

If the plant belongs to a national or a foreign company it makes a difference with respect to environmental progress. In other words, the findings of the dynamic assessment differ on this point, but not on the other points, from those of the static assessment. This might imply some consistency in the recent context of increasing liberalisation with higher international pressures, domestic difficulty – particularly in the case of small to medium size firms – and the lack of technical environmental information.

Conclusions

Increasing liberalisation through NAFTA induced three important effects in the steel sector. First, the scale of production increased, growing at higher rates than most manufacturing industries in Mexico. Second, the composition of exports changed, increasing the relative importance of finished and final-consumption iron and steel products, with respect to primary goods. Third, technological change occurred, reflected in growing productivity and the adoption of modern equipment. Furthermore, in the early years of NAFTA, the steel sector reduced its energy intensity. Accordingly, there is evidence indicating some progress in environmental management and performance by this sector. However, such progress is confined to a few firms and does not characterise the Mexican steel sector as a whole. There is still a lot to do in the case of a significant number of plants, according to our assessment.

Although it is not possible to generalise, since this assessment is based on a sample interview, which is not random, these findings provide a rough idea of the steel sector at plant level. According to our static analysis, there seems to be a large dispersion among steel plants with respect to their environmental policies, actions and performance. In 1998, it appears that the main emphasis was placed on private policy design rather than on effective implementation. Export-oriented plants have more developed environmental management, and this is associated with different access to technological options and with different market incentives. Moreover, the largest national plants (not the subsidiaries of multinational firms) tend to adopt cleaner technologies more rapidly than the smaller national plants, and tend to comply with the local standards (and even perform better than the official standards require) mainly because of financial and technological factors.

It seems to be that the size, market orientation, technological level and the type of process (in terms of vertical integration) play a central role in a firm's behaviour regarding industrial pollution at a point in time. The *size* of a plant can be associated with important economies of scale and scope in pollution control and with advantages in financial, human and organisational resources. The *market orientation* of production (particularly to foreign markets) suggests the importance of standards of quality and environmental protection. The *technological level* is possibly associated with a favourable technological culture and access to restricted technical information concerning pollution control. Finally, the *type of process* implies the central role of greater environmental hazards and greater intensity of pollution of primary processes, such as steel making, which require more attention to be paid to the environment.

Two interesting motivations were identified. First, although government regulation is the main motive for better environmental behaviour, foreign market pressure is a more important factor amongst firms that perform well than for those who do badly. Second, government environmental regulation is a more important motive for poor performers than for good performers. The high equipment costs, the lack of a supply of 'cleaner' technologies, the need to attend to other more urgent priorities, and a lack of incentives and of technical and environmental information are the main external and internal obstacles for firms that are lagging behind in their environmental performance.

The dynamic analysis reveals an environmental assessment lower than that of the static evaluation. Furthermore, it suggests greater heterogeneity of environmental trends, with cases of plants with a high rating, and other plants with very low ratings. This dispersion is even higher than that related to the static evaluation.

The size of steel plants, technological level and type of process seem to be the key factors inducing different environmental evolution. The market orientation, in terms of the volume of exports, is also a central variable. In particular, large plants, with significant exports, are the best performers in terms of progress in environmental management. Vertically integrated plants have the best assessment in relation to the evolution of environmental performance. The plants with high technological level reached the highest score for productive changes with environmental purposes, and also had a high assessment concerning both the evolution of environmental performance and improvements in environmental management.

Both the static and the dynamic analyses indicate that the group of poorly evaluated plants face important obstacles to better environmental performance. First, internal problems of inefficiency and debt lead firms to attend to priorities other than environmental problems. Second, there is a lack of economic incentives that would make it cheaper to reduce pollution and more expensive to continue to contaminate. Mexican public policy is playing an important role through environmental standards, but lacks economic mechanisms. The introduction of economic instruments (either fiscal or financial instruments or both) might be an appropriate way of dealing with these problems.

Notes

1 I wish to express my appreciation to Rhys Jenkins for his excellent research supervision, and for his valuable comments and careful revision to a previous draft. I am also grateful to Lourdes Aduna, Flor Brown, Lilia Domínguez, Oscar Fernández, Francisco Giner, Horacio Sobarzo and Victor Urquidi for their critical comments and useful suggestions in two seminars held at El Colegio de Mexico and Instituto Nacional de Ecología, in Mexico City, in 1998. Thanks to Nuyavi Malpica for her good assistance in this research. Finally, I would like to thank environmental managers at 12 Mexican steel plants for their co-operation; this study depended on the information provided by them.

2 Although high and medium exporters perform better than low exporters, the best performance is found amongst the medium exporters.

3 Integrated firms seem to be obliged to take such measures due to the fact that the primary process of steel making is particularly pollution intensive.

Bibliography

Barton, J. R. (1999) *Environmental Regulations, Globalisation of Production and Technological Change in the Iron and Steel Sector*. School of Development Studies, University of East Anglia, Norwich, UK (draft working paper).

Brown, F. and A. Guzmán (1998) 'Cambio Tecnológico y Productividad en la Siderurgia Mexicana, 1984–1994', *Comercio Exterior*, 48 (10): 836–44.

Canela, J. A., H. Cárdenas and A. Guevara (1998) 'Los Sistemas de Información Computarizados como Herramienta para Mejorar la Aplicación de la Ley Ambiental: El Caso de la Dirección General Jurídica de la PROFEPA', *Comercio Exterior*, 48 (12): 987–94.

Grossman, G. and A. Krueger (1992) *Environmental Impacts of a North American Free Trade Agreement*. Washington, DC: National Bureau of Economic Research, Working Paper No. 3914.

Hettige, H., P. Martin, M.Singh and D. Wheeler (1995) *The Industrial Pollution Projection System*. Washington, DC: World Bank, Policy Research Working Paper 1431.

IISI (1999) *Steel Statistical Yearbook 1998*. Brussels: International Iron and Steel Institute.

INE (1996) *Instrumentos Económicos y Medio Ambiente*. México, DF: Instituto Nacional de Ecología.

INEGI (1994) *XIV Censo Industrial. Industrias Manufactureras, Extractivas y Electricidad*. México, DF: Instituto Nacional de Estadística, Geografía e Informática.

INEGI (1997, 1998, 1999a) *La Industria Siderúrgica en México*. México, DF: Instituto Nacional de Estadística, Geografía e Informática ('Editions 1996, 1997 and 1998').

INEGI (1999b) *El Sector Energético en México. Edition 1998*. México, DF: Instituto Nacional de Estadística, Geografía e Informática.

Mercado, A. (ed.) (1999) *Instrumentos Económicos para un Comportamiento Empresarial Favorable al Ambiente en México*. México, DF: El Colegio de México/Fondo de Cultura Económica.

Mercado, A. and O. Fernández (1998) 'La Contaminación y las Pequeñas Industrias en México,' *Comercio Exterior*, 48 (12): 960–5.

Mercado, A., L. Domínguez and O. Fernández (1995) 'Contaminación Industrial en la Zona Metropolitana de la Ciudad de México', *Comercio Exterior*, 45 (10): 766–74.

Appendix 1: Methodology used for evaluating firm environmental performance

Performance at a point in time

The evaluation of performance at a point in time involved assessment at both the corporate and plant level. At the level of the firm, it is necessary to assess the type of policy that exists for environmental protection, the actions taken in the corporation for that protection, environmental management, and the degree of environmental control achieved. At the level of the manufacturing plant, it is necessary to analyse the series of actions based on the plant's environmental programmes and environmental performance. On the basis of these two levels, it is possible to arrive at an overall assessment of the firm in question. Those firms with a high score may be considered as successful in terms of environmental behaviour and performance. Firms with low scores lag behind in their environmental performance and behaviour. More specifically, the static assessment follows the criteria summarised in Table 10.A1. The maximum possible score for each item was taken as a maximum index equal to 10, and the actual scores reached by each plant were calculated as a proportion.

Table 10.A1 Environmental assessment criteria at a particular point in time

Assessment aspects	*Maximum no. of points*
A. Corporate level	19.5
A.1 Environmental policy	4
A.2 Actions with programmes	6
A.3 Environmental management	7.5
A.4 Firm's environmental performance	2
B. Plant level (average)	11
B.1 Actions with programmes	4
B.2 Plant's environmental performance	7
C. Overall (average)	30.5

Dynamic performance

To evaluate the performance of a firm over time, attention was focused on changes that have been made in production and environmental management, which are designed to reduce the environmental impact of the firm. These elements were grouped into three issues: (a) productive changes for environmental purposes, (b) the evolution of the plant's environmental performance, and (c) improvements in environmental management. The sum of all the assigned points under these headings gives an overall environmental assessment of the plant in question over time. As in the static evaluation, firms with a high score are considered to be making significant environmental improvements while those with a low score are doing little to reduce their environmental impact over time. The dynamic assessment follows the criteria summarised in Table 10.A2. The maximum possible score for each item was taken as a maximum index equal to 10, and the actual scores reached by each plant were calculated as a proportion.

Appendix 2: The sample of firms surveyed

As was explained above, this research focuses on the most polluting steel processes, and other downstream processes for manufacturing finished steel products. We therefore decided to study two groups of firms: (1) the major steel companies with steel making and rolling, and (2) a variety of companies that carry out any of the following processes: steel making, rolling and product finishing, particularly tubes and structural pipes and sections. According to the National Chamber of Iron and Steel Industry in Mexico ('Canacero'), there were 46 plants in these two groups of process in Mexico, in 1997, when this research was starting.

Table 10.A2 Environmental assessment criteria over time

Assessment aspects	*Maximum no. of points*
A Productive changes for environmental purposes	4
Technological changes in equipment	1
Substitution of less polluting inputs	1
Changes in the productive process	1
Changes in the product mix	1
B Evolution of environmental performance	6
B.1 In terms of attention paid to volume of resources used	4
Less intensive use of fuel	1
Less intensive use of electricity	1
Less intensive use of water	1
Lower volume of water discharged	1
B.2 In terms of attention paid to quality of emissions	2
Higher quality of water discharged	1
Higher quality of air emissions	1
C Improvements in environmental management	1
D. Overall (average)	11

At the outset of the research, an objective had been set involving a sample size of 20 plants, taking into account that the statistical universe includes 46. Unfortunately, the degree of refusal was so high that it was only possible to arrive at a total of 12 firms. However, this number of firms represents 26 per cent of the total (Table 10.A3). We should also bear in mind that it includes Mexico's major steel plants (and a variety of other firms according to the criteria established), so that in terms of employment, the group of firms surveyed (Table 10.A4) represents 54 per cent of total employment in the statistical universe. In this sense, the sample allows us to arrive at a fairly

Table 10.A3 The sample and the statistical universe, 1997–8

Concept	*No. of firms*	*Employment (persons)*
Statistical universe	46	41,768[a]
Survey	12	22,469
Share of survey in the statistical universe	26.1%	53.8%

Note

a Data from 1996. The total number of persons employed by the group of firms surveyed was 23,980.

Table 10.A4 List of plants included in the survey interview, 1998.

1	AMSCO Mexicana, SA
2	Altos Hornos de México, SA (AHMSA)
3	Cía. Mexicana de Perfiles y Tubos, SA
4	Cía. Acerera de California, SA (CSC)
5	Cía. Acerera de Guadalajara, SA (CSG)
6	Aceros CORSA, SA
7	Hojalata y Lámina, SA (HYLSA)
8	Ispat Mexicana, SA (IMEXSA)
9	Mexicana de Laminación, SA
10	Productos Laminados de Monterrey, SA (PROLAMSA)
11	Acerera Lázaro Cárdenas Las Truchas, SA (SICARTSA)
12	Tubos de Acero de México, SA (TAMSA)

good picture of the industry, with a certain degree of reliability, although strictly speaking it does not comprise a good statistical base for making generalisations.

11
CONCLUSION

Rhys Jenkins

This volume has provided a wealth of empirical material on the much neglected issue of the environmental impact of industrial production in Latin America. Inevitably in a new area of research of this kind, there are many gaps in our knowledge of the field. A collection such as this can throw light on some of the issues and highlight the areas in which further research is needed. By way of conclusion, therefore, this chapter discusses the current state of knowledge on some of the questions which were touched upon in the introduction, drawing on the evidence that has been presented in the various studies.

At the outset it should be noted that empirical research on industrial pollution in Latin America is severely hampered by the almost complete lack of good data. At the national level, measures of industrial emissions are totally lacking. Researchers are, therefore, forced to rely on estimates, based on coefficients taken from developed countries, particularly the USA. Such estimates do not take into account differences in technology and in pollution abatement efforts between the developed countries and Latin America. Moreover, since they are based on fixed coefficients, their use over time does not reflect any shifts towards greater eco-efficiency amongst firms in the region. As such they provide only the crudest estimates of industrial pollution in the Latin American countries.

Another problem is the difficulty of obtaining data at the firm level. There are a number of issues here. First, there is the sensitivity of environmental issues as far as many firms are concerned. While a few are keen to project a 'green' image and take pains to publicise their environmental efforts and achievements, many feel that the environment is a stick which can be used to beat them. As a result they are reluctant to divulge information about their environmental performance or to give access to researchers. This is reflected in some of the studies published here such as those on textiles and iron and steel in Mexico where many firms refused to be interviewed about their environmental activities. In other cases, where firms have only recently become more environmentally aware, they lack past data on environmental performance simply because this was not previously monitored. Moreover, many firms, particularly small and medium enterprises, still do not monitor their own performance in any systematic way.

The lack of centrally collected data on the one hand, and the non-availability of firm-level data on the other, presents particular problems for the kind of sectoral analysis which has been central to this book. Researchers can neither resort to aggregate data on the emissions of the industry as a whole, nor can they easily obtain figures for the emissions of the major firms.

Environmental performance in Latin American industry

Although the lack of data makes it impossible to arrive at a comprehensive evaluation of environmental performance in Latin American industry, there is clear evidence that during the 1990s the business sector began to recognise environmental issues as important and that firms have taken steps to improve their environmental performance. Thus, comparing the situation at the start of the twenty-first century with that of a decade earlier provides a positive picture of the progress that has been made. It is useful to discuss corporate environmental performance under four main headings:

1 Adoption of environmental policies
2 Implementation of environmental management systems
3 Environmental programmes and activities
4 Reductions in environmental impacts

The first three of these are, in practice, indicators of inputs, while only the fourth can be considered an output indicator of environmental performance.

Environmental policies

A number of the larger firms in the region have adopted explicit environmental policies and some also participate in local chapters of the Business Council for Sustainable Development or have signed up for programmes such as the Responsible Care programme of the chemical industry. Chudnovsky *et al.* observe that such policies were common amongst the large Argentinian firms which they surveyed. In Mexico a number of firms in the synthetic fibre and iron and steel industries had adopted environmental policies, although this was less common in the textile industry. Similarly in Brazil, some of the large iron and steel and pulp companies have explicit policies and documents which set out their environmental achievements. However, in the leather industry such policies do not appear to be widespread.

Of course, even where firms do have an environmental policy, there may be considerable differences in how developed and comprehensive such policies are in terms of their coverage. Thus, for some firms a policy could be mainly to seek to comply with local environmental norms, while others may include more far-reaching commitments to continuous improvement or meeting international corporate standards. There is obviously scope here for more detailed evaluation of the nature of firm environmental policies.

Environmental management

As was indicated in Chapter 1, there has been a rapid growth of environmental management in Latin America in the second half of the 1990s. One indication of this is the increasing number of firms certified under ISO 14000 (see Table 1.2). However, at the end of the 1990s, there were only just over 200 certified firms in the region, the vast majority of them in Argentina, Brazil and Mexico. Not surprisingly in view of the relatively small number of firms which have already been certified, only a few of the firms discussed in the various chapters in this book have achieved ISO 14000 certification, although there is evidence that a larger number are intending to obtain it and have taken some steps in that direction.

Independently of ISO 14000, a number of the larger firms in the three countries have introduced some form of environmental management. In Mexico, for instance, 70 per cent of large firms in a survey reported having formal environmental management systems (Lexington, 1996: Fig. IV.1). In some cases this has involved setting up a separate environmental department with a responsible manager, although often these functions are combined with other aspects such as health and safety or production.

Environmental programmes and activities

A number of the firms looked at in the case studies have environmental programmes of one kind or another. These can range from training for their own staff to involvement with local communities. However, although almost all the large firms surveyed in Argentina had some form of environmental training, it was not clear how substantial these were, either in terms of the amount of training received or the number of employees involved. In Mexico, there is evidence that most of the chemical fibre companies provide some environmental training, although again the amount of training provided is not known. In addition to such activities, firms also have programmes to reduce emissions and wastewater discharges, to recycle waste and to reduce energy use.

Not only are firms establishing various kinds of environmental programmes, but they are also increasingly undertaking environmental investments. As was noted in Chapter 1, the market for environmental equipment in Latin America is growing rapidly. Compared to their previous record, a number of firms significantly increased their environmental investments in the 1990s. Amongst large Argentinian firms the share of environmental investment in total investment almost doubled from 10 per cent in 1993 to 19 per cent in 1997 (Table 4.1). A survey of Mexican firms in the mid-1990s also found that firms were increasing their environmental expenditures (AmCham, 1993). The Brazilian case studies of iron and steel companies and the pulp exporters also indicate that they sharply increased their environmental investments during the 1990s.

Environmental impact

The critical question ultimately is not the extent to which firms adopt environmental policies or introduce environmental management systems, but rather how far they are able to reduce their negative impact on the environment. It is of course generally assumed that changes in policy and management will lead to improvements in environmental performance, but this need not always be the case.[1] It is therefore essential, wherever possible, to consider the actual environmental performance of firms in terms of their use of resources and level of emissions.

The information in the case studies on the environmental impact of the firms is at best fragmentary. This reflects the problems of data availability discussed earlier. However, there are clear examples of firms which have achieved substantial improvements in their performance in terms of a number of environmental indicators. In Brazil for instance, Aracruz in the pulp industry reduced its emissions dramatically during the 1990s (see Fig. 7.5) as did CST in the iron and steel industry (see Chapter 5). A number of chemical fibre plants in Mexico also claim to have achieved substantial reductions in their emissions and water discharges, and improvements in the quality of wastewater (see Table 9.9).

Another performance indicator which has improved is energy efficiency. Chudnovsky *et al.* report significant advances by large Argentinian firms in this area. Similarly, data on energy consumption per tonne of steel produced indicate that the iron and steel industry has succeeded in increasing energy efficiency in both Brazil and Mexico during the 1990s (Fig. 5.7 and Table 10.2).[2] Improvements in energy efficiency were also observed in the synthetic fibres and textile industries in Mexico.

Limits of corporate environmental performance in Latin America

While there is clear evidence of progress in corporate environmental performance in Latin America since the early 1990s, it is also important to note several limitations in terms of what has been achieved.

First, it is clear that this progress has been highly uneven. It is possible to point to examples of best practice in environmental management and significant reductions in pollution indicators, but these are by no means generalised across all industries or even within particular sectors. Several of the case studies point to the extent of intra-firm variation in terms both of environmental management and impacts. Thus, the scope of environmental management amongst large Argentinian firms is observed by Chudnovsky *et al.* to be highly uneven. In Mexico both the textile industry and the steel industry show considerable variation in environmental performance at the firm level, while the same is true in the leather and the iron and steel industries in Brazil. Of the sectors studied here, only the synthetic fibres industry in Mexico and the pulp export industry in Brazil are characterised by a more or less general pattern of environmental improvement, although even in these two industries there are clearly some firms which are more advanced than others.[3]

One of the factors which accounts for the differences in environmental performance within industries is the tendency of smaller firms to lag behind in terms of both the development of environmental management and of programmes to reduce emissions and resource use. As was noted in Chapter 1, previous studies from Mexico have found that managerial effort in relation to the environment is positively related to firm size and that small firms tend to pollute more relative to their size than larger firms. There are clearly economies of scope in environmental management, and small firms do not have the resources to develop specialised environmental departments. Equally there are economies of scale in terms of pollution abatement, which make the cost of reducing emissions proportionately greater for small firms.

Several of the studies in this book confirm these perceptions. Chudnovsky *et al.* (1996) demonstrate that environmental management and pollution prevention are much more advanced amongst the sample of large Argentinian firms than in the small and medium enterprises of Greater Buenos Aires. In all three of the Mexican case studies, environmental performance was positively related to the size of firm. This was true even in the case of synthetic fibres where, although there were no SMEs, the largest firms (those belonging to the top 100 firms in the country) performed better than the others. Moreover, these differences were consistent across the whole range of environmental indicators, both at the corporate and the plant level. In Brazil, Odegaard comments on the poor environmental performance of small tanneries producing mainly for the domestic market. In neither the pulp industry nor the iron and steel industry in Brazil was the size issue looked at, since the firms covered were all large. It is interesting to note, however, that in the case of the iron industry, small-scale pig iron producers (*guseiros)* are a major source of environmental problems. Other studies of corporate environmentalism in Brazil have also noted a positive correlation with firm size – see Roberts and Stauffer (2000) on the Brazilian chemical industry.

Although it is undeniable that some progress has been made by the corporate sector, performance in Latin America lags a long way behind that achieved in North America and Western Europe. The case studies provide a number of illustrations of this gap. In the Brazilian leather industry, for instance, environmental control costs are much lower than in Italy, reflecting a lack of emphasis on pollution abatement (see Table 6.3). In the steel industry in Brazil, despite the growth of environmental investment in recent years, the share of total investment is lower than for steel plants in Europe.[4] Similarly, levels of recycling/resale of wastes and of energy efficiency are lower than in the steel industry in developed countries (see Chapter 5). Even in the pulp industry, where Dalcomuni reports a relatively good environmental performance (see Table 7.3), Carrere (forthcoming) finds that environmental investment levels are lower than those achieved in industrialised countries.

Another limitation is the nature of the environmental progress that has been achieved so far in the region. First, there is clear evidence that 'end-of-pipe' solutions tend to dominate and that relatively little has been achieved in terms of introducing 'clean technologies'. In so far as firms have adopted pollution prevention, as opposed to 'cleaning up' measures then these have tended to be of a relatively 'simple' variety.[5]

The sectoral studies show that in the steel, textile and leather industries, the focus has tended to be on 'end-of-pipe' measures. In the case of textiles and leather, this usually involves some form of wastewater treatment. In the steel industry, attention has concentrated on atmospheric emissions and solid wastes. Emissions have been reduced through the use of bag filters and electrostatic precipitators, and waste has been processed so that it can be reused or resold. The increasing importance of cleaner processes such as DRI, EAFs and continuous casting, particularly in Mexico (see Chapter 10) and Argentina (Chudnovsky *et al.*, 1996: Ch. VIII), have not been introduced for environmental reasons, but rather reflect the growing competitiveness of these technologies.

There is more evidence of pollution prevention playing a significant role in the Mexican synthetic fibres industry and the Brazilian pulp industry. In the Mexican case, Dominguez reports that three of the six firms regarded environmental management as the main source of environmental improvements and two firms mentioned process modification as the major factor.[6] Recycling and treatment were mentioned as the second factor. This suggests that in this industry, at least, firms have gone beyond merely an 'end-of-pipe' approach, although it is not clear whether the pollution prevention measures go beyond relatively simple ones. The pulp industry has gone furthest of any of the case studies in this book in developing more complex forms of pollution prevention and particularly introducing cleaner production methods to reduce chlorine emissions.

Sectoral patterns of environmental performance

As was noted above, the development of corporate environmentalism is uneven, not only as between different firms, but also across industries. The industries included in this volume, together with sectoral studies published elsewhere, give an overview of those industries which account for much of the region's industrial pollution: mining, metallurgical industries, pulp and paper, chemicals, leather tanning and textiles.

Mining

By its very nature, mining has a major impact on the local environment and historically this has been highly damaging in terms of land degradation, atmospheric pollution and effluent discharges. Warhurst and Hughes-Witcomb present a very positive picture of environmental developments in Latin American mining in recent years. They illustrate this with examples of new cleaner technologies which have been introduced in mineral extraction in the industry, and the implementation of environmental management.

While there is no doubt that improvements are being made, Chapter 3 concentrates particularly on large-scale projects often involving high-profile TNCs and international financial institutions. Although some mining is large scale and capital intensive, there are also numerous examples of highly polluting, small-scale mining. The activities of *garimpeiros* in gold mining in Brazil, which have resulted in major problems associated with mercury pollution, are a notorious example. Nor are environmental problems in the region confined to small-scale producers. Runge *et al.* (1997) cite the example of Peru where tailings and mine wastes are largely unregulated and have affected drinking water and marine life along the Peruvian coast.

Pulp and paper

The pulp and paper industry and the chemical industry have been the most exposed internationally to environmental pressures and it is not surprising that they are amongst the most environmentally active sectors in Latin America. As Dalcomuni makes clear, international pulp producers have had to respond to the pressures from environmental NGOs to eliminate chlorine from their products. This has required significant process changes in order to reduce emissions. As a result, large firms in the industry have taken a lead, not only in adopting cleaner technologies, but also in adopting environmental policies and introducing environmental management systems. This has been the case not only in Brazil but also in other Latin American countries where firms are seeking to supply the international market (see Chudnovsky *et al.*, 1996: Ch.VI on Argentina, and Scholz *et al.*, 1994 on Chile).[7]

Chemicals

The chemical industry is also environmentally highly sensitive because it uses and produces toxic and hazardous materials and has been the cause of a number of major incidents internationally and in Latin America. The chemical industry is rather under-represented in this volume with only one study and that of a relatively small part of the industry, synthetic fibres. However, other studies in Argentina and Brazil confirm that the chemical industry is also one in which environmental issues are being given increasing weight (see Chudnovsky *et al.*, 1996: Ch. VII on the Argentine petrochemicals industry, and Roberts and Stauffer, 2000 on the Brazilian chemical industry).

Environmental management is relatively more developed in the chemical industry than in other sectors.[8] National chemical associations in the region have followed Canada and the USA in adopting Responsible Care programmes that set certain standards for their members. Large national and multinational firms dominate the most polluting sectors of the chemical industry. These are highly visible and concerned to avoid getting a reputation for environmental irresponsibility. The Brazilian chemical association ABIQUIM now requires its members to participate in the industry's *Atuação Responsavel* programme (Roberts and Stauffer, 2000), while the Mexican association ANIQ has its own *Responsabilidad Integral* programme in which, as Dominguez mentions, a number of the synthetic fibre companies have been involved.

Iron and steel

The iron and steel industry, although also a major polluter, has not been subject to the same environmental pressures as either the pulp and paper industry or the chemical industry internationally. It is not surprising, therefore, that the industry in Latin America has not been as proactive in terms of environmental measures as sectors of the chemical industry or the pulp and paper industry.

A major change in the industry in all three Latin American countries covered here has been the privatisation of the state-owned firms which previously dominated the industry. In the past they often lacked the resources to carry out necessary environmental investments, and it has been suggested that because they were state owned, the environmental authorities were reluctant to act against them. This has changed with privatisation and as Barton shows in his chapter, there has been a surge of new investment in the Brazilian steel industry which has included investments in environmental protection. However, this continues to be mainly of the 'end-of-pipe' variety.

Traditional industries

Both leather tanning and textiles are traditional sectors with significant numbers of SMEs. As traditional industries they often have old plants, in some cases established up to a century ago. The most pressing environmental problems for both industries is the treatment of wastewater. In the case of tanning, there is a particular problem involving the use of chromium in the production process. This problem was highlighted in the mid-1990s when large numbers of migratory birds from Canada died in Mexico near the tanning industry centre of León, allegedly from chromium-containing effluents (Knutsen and Wik, 1999). As a result, measures were introduced to try and reduce pollution by the leather industry.

Of the industries covered in this book, these are the ones where the lack of environmental measures and concern are most apparent. Many of the firms have little if any awareness of their environmental impact. They do not monitor their emissions, they lack environmental management systems or policies, and they often work with outdated plant and equipment which generate high levels of pollution, without having any wastewater treatment facilities. Even in these industries there are examples of firms which are proactive on environmental matters, but these are the exception rather than the rule.

Drivers of corporate environmentalism

Given that there have been some undeniable improvements in corporate environmental performance, however partial and uneven, in Latin America over the past decade, the question that must now be asked is what factors have been responsible for promoting such a change. Is it primarily a response to government regulation? How important a role have local community pressures played? Have market forces helped to push firms in this direction?

There is no doubt that environmental issues have moved up the policy agenda in Argentina, Brazil and Mexico over the past decade or so (see Chapter 1). However, most observers also agree that environmental regulation remains uneven and inadequate. Government environmental agencies are underfunded and lack the human resources to inspect the plants for which they are responsible. There are often problems of competing jurisdictions and lack of clear responsibilities between national, regional and local authorities. Particularly in Brazil, where the states play a major role in environmental regulation, there are considerable differences in the stringency of regulation and enforcement in different regions. Moreover, the recent trend towards reducing the role of the state in Latin America means that increasing environmental regulation swims against the tide of deregulation and cuts in government expenditure which has been in full flow since the mid-1980s.

Under these circumstances, has government regulation played a significant role in pressurising firms to improve their environmental performance? The three Mexican case studies in this volume all find that regulation has been the main driver to environmental improvements. This confirms the findings of other studies of Mexican firms. A report for the World Bank found that over 60 per cent of the firms interviewed considered regulatory requirements a strong or very strong influence on their environmental actions (Lexington, 1996: Fig. II.1). Dominguez (1999: Table IV.13) also found that regulation was mentioned more often than any other factor as a motive for environmental improvements.

The studies of iron and steel and of tanning in Brazil also point to the importance of regulation in motivating firms to deal with environmental problems. In the case of the leather tanning industry, however, it is noted that enforcement is often weak and the evidence from iron and steel also suggests that it is highly uneven. In the case of the pulp industry it is noted that regulation was an important factor up to the 1980s, but that, since then, the emphasis has shifted. This change was also associated with a shift from 'end-of-pipe' measures to pollution prevention as a major means of improving environmental performance on the part of the firms.

The study of Argentina shows that regulation is not the major factor leading firms to adopt *pollution prevention*. In fact it is ranked fourth which, as Chudnovsky *et al.* comment, is not surprising since regulation is based on 'command and control'. Since the chapter focuses primarily on pollution prevention, it does not provide any indication of the impact of regulation on other aspects of environmental performance. Other studies of Argentina, however, confirm that regulation has played an important role in the past. In a survey carried out by the Consejo Empresario para el Desarrollo Sostenible and the Asociación para el Desarrollo de la Gestión Ambiental in the mid-1990s, 97 per cent of the firms covered indicated that they were influenced by local legislation (quoted in Chudnovsky and Chidiak, 1995). Similarly, sectoral studies of the Argentine petrochemicals and steel industries also point to the crucial role played by stricter regulatory requirements (Chudnovsky *et al.*, 1996: Chs VII and VIII). In contrast, regulation seems to have been a less significant factor in the pulp and paper industry (Chudnovsky *et al.*, 1996: Ch.VI).

In summary, therefore, it is clear that environmental regulation has been a key factor driving firms to improve their performance. Because regulation has been largely based on command and control, it has been more successful in encouraging firms to adopt 'end-of-pipe' measures than in bringing about the introduction of cleaner technologies. It has, however, been critical in getting firms to begin to take environmental issues seriously. While regulation was pretty well non-existent or not enforced, firms were happy to externalise environmental costs and faced no incentive either to invest in pollution abatement or to develop environmental management systems. Subsequently, some firms go beyond what is required by regulation or find that they can reduce the cost of compliance by various strategies of pollution prevention. There is no doubt, however, that, in the absence of efforts at regulation, few if any firms would have got to this point.

In some cases, firms improve their environmental performance in response to pressure from neighbouring communities which are affected by pollution. There are certainly specific examples of such pressures. Odegaard for instance mentions the role played by the local community in Estancia Velha, together with more stringent regulation by the state authorities in Rio Grande do Sul, in forcing the local tanning industry to clean up in the 1980s. Firms in the Mexican textiles and chemical fibre industries identified local community concerns as a factor in their environmental decision making, and in a general survey of firms about a quarter cited local community pressure as a strong or very strong influence (Lexington, 1996: Fig. II.1)

One of the best known examples of community pressure around industrial pollution in Latin America was in Cubatão. Faced with the urban pollution that led to the town being described as the most polluted in the world and referred to as the 'valley of death', the Association of the Victims of Pollution and Bad Living Conditions (AVMP) was set up in the early 1980s. The AVMP played an important role in demanding that local industry reduce its pollution levels and resulted in substantial improvements in air quality (de Mello Lemos, 1998).

Another potential driver for environmental improvement is a firm's customers. This is a consideration for some firms and in some industries. As Dalcomuni shows, the requirements of customers, particularly in the EU, have been the key factor in promoting environmental change in the Brazilian pulp industry. There is also evidence that this has been an important factor for one major pulp exporter in Argentina (see Chudnovsky *et al.*, 1996: Ch. VI). However, these findings cannot be generalised. The study of large Argentinian firms in this volume does not find that market demands are an important driver, even in the case of exporting firms, and a previous Argentine study found that export market requirements were a motivating factor for only 17 per cent of the firms studied (quoted in Chudnovsky and Chidiak, 1995).

The three case studies of Mexico show that clients play a limited role in motivating firms to improve their environmental performance. In the case of textiles it is an insignificant factor (see Fig. 8.3), while for a few firms in chemical fibres and iron and steel, it plays a minor role. This is again in line with the findings of a previous Mexican study where only one-fifth of the firms regarded customers as a strong or very strong influence (Lexington, 1996: Fig. II.1).

In so far as market demands do influence environmental performance, it is almost exclusively related to the demands of customers overseas. The environmental performance of firms according to their export orientation is discussed in several of the case studies, and the evidence has already been summarised in Chapter 2. The picture that emerged is rather mixed. While there is some evidence to support the view that exporters perform better than firms which produce for the domestic market, there are clear inter-industry differences which are consistent with the finding that the importance of customer pressure also differs from industry to industry.

Another external factor, which has led to some firms introducing environmental measures, has been the requirements of international financial institutions. Warhurst and Hughes-Witcomb highlight the importance of this as a factor in the mining industry in Latin America. In Bolivia, for example, they report that mining executives considered the conditions imposed by international lenders as more important in influencing environmental behaviour, than national regulation. In contrast, Chudnovsky *et al.* find that this has not been an important driver amongst large Argentine firms, and it is not mentioned in any of the manufacturing sector case studies.

A number of other factors, which are internal to the firm, may also play a part in motivating business to improve its environmental performance. Some firms regard it as an important part of their public image, not wishing to be seen as environmentally irresponsible. The study of Argentine firms found this to be the second most important factor promoting pollution prevention amongst large firms, and the most important factor for foreign-owned firms. This is not surprising since the importance attached to public image depends critically on the extent of the firm's profile, and it is likely to be particularly important to TNCs who find their overseas activities subjected to the glare of publicity by environmental NGOs and the media in their home countries. Protection of reputation is, therefore, a particular concern for such firms.

Although public image has not been the major driver behind the recent environmental measures taken in the pulp industry in Brazil, the major firms are clearly concerned to project a 'green' image, particularly in relation to their plantation activities. They are at pains, for instance, to emphasise the sustainability of their practices and the fact that they do not contribute to deforestation in the Amazon (Carrere, forthcoming). The other Brazilian case studies illustrate the importance of commitment by a firm's top directors as a factor in environmental performance as in the case of Usiminas which became the first firm to achieve ISO 14000 certification.

In Mexico only a minority of firms in the textile, synthetic fibre and iron and steel industries identified public image as an important factor (Fig. 8.3; Tables 9.4 and 10.14). A somewhat more positive picture was found in a study of firms in Mexico City where almost a quarter identified public image as the most important factor, prompting them to invest in environmental protection, and a further quarter ranked it as the second factor (Dominguez, 1999: Table IV.13).

Several comments on the role of public image need to be made. First, in so far as what is being promoted is an *image*, a firm may find cost-effective ways of doing this which do not represent substantial environmental improvements. In other words there is a danger that the response will be some form of 'greenwashing' by the firms involved. As Carrere (forthcoming) points out for the case of the Brazilian pulp industry, some of the claims of environmental responsibility made by the firms do not stand up to careful scrutiny.

A second limitation to the role of public image is that it is only really relevant in the case of firms which have a high profile within the country or internationally. Small or medium enterprises, and even large firms in industries which do not draw much public attention, are unlikely to be motivated to improve their environmental performance by such considerations.

A further point is that reliance on a particularly committed top manager to bring about environmental actions inevitably means that these will be uneven and cannot be generalised to the corporate sector as a whole.

So far we have considered factors which directly encourage firms to adopt environmental measures. However, it is important to note that environmental improvements may also come about as an indirect result of changes which are motivated by other factors, in particular by technological modernisation and attempts to reduce input and utility costs.

There are numerous examples of the way in which technological changes have tended, as a by-product, to reduce industrial pollution. Amongst large Argentine firms, measures to reduce costs were the third most important factor leading to the adoption of pollution prevention measure (see Chapter 4). In the iron and steel industry, particularly in Mexico and Argentina, the increasing significance of DRI, EAFs and continuous casting, none of which have been adopted for environmental reasons, have been important factors in reducing emissions. In mining, cleaner technologies are only adopted when they also reduce costs, and in many cases the environmental advantages are incidental to the main motive which is cost saving (see Chapter 3). More generally, newer plants tend to be less polluting and more efficient than older ones, thus reducing both emissions and resource and energy use.

The corollary of this is that where an industry is in economic difficulties and investment levels and profitability are low, then significant environmental improvements are less likely. The Mexican textile industry in recent years has been a case in point, where there has been very little new investment and hence little technological or environmental upgrading.

Efforts to reduce costs have also contributed to environmental improvements. Chudnovsky *et al.* (1996: 313) cite the example of the paper industry in Argentina where a significant increase in the reuse of trimmings led to a substantial reduction in the use of pulp. Also, in a number of the case studies, energy saving measures and conversion to natural gas have helped reduce pollution.

This review highlights the significance of regulation, however inadequate and uneven it currently is in Latin America, as an important factor in all the industries which have been studied here. It is the one factor which can be said to have a universal effect in promoting better environmental performance, at least where enforcement is effective. A number of other factors are influential in some industries or for some firms. Thus, for instance, some exporters are responding to pressures, or anticipated pressures, from their customers. This is most marked where regulation overseas makes it essential to respond in order to secure market access. Thus, European regulations on the chlorine content of pulp has meant that this has been a key sector in which the demands of clients overseas has led to environmental improvements. On a much smaller scale, some leather exporters from Brazil and Argentina have also responded to German restrictions on the use of certain dyes and chemicals (see Chapter 6 and Chudnovsky and Chidiak, 1995).[9]

The evidence that local communities and public image are important drivers of environmental improvement is less widespread. The former tend to be confined to particular communities which have suffered from serious pollution problems and have achieved a certain level of organisation. The latter may not lead to substantial improvements in environmental performance and can result merely in green window-dressing. Even if substantive improvements do occur, they are likely to be limited to highly visible firms and not to be generalised across the industrial sector.

Finally, it is important to underline that many environmental improvements may come about indirectly as a result of measures which are designed to increase profitability and are not motivated by environmental considerations at all. In such a case it is inappropriate to speak of corporate environmentalism since firms are merely responding to profit opportunities and seeking to reduce costs, as any firm will want to do.

Obstacles to improving environmental performance

Many of the studies in this book throw light on the obstacles which prevent firms improving their environmental performance or limit the extent of the improvements which they are able to achieve.

The three Mexican case studies explicitly questioned firms concerning the internal and external (to the firm) obstacles which they face regarding environmental performance. Although when questioned directly about environmental issues, managers often claim that they are given a high priority, the most common internal obstacle cited by Mexican firms, in all three industries, is that they have other priorities. Thus, despite the advances that have been made, it is clear that the common perception of environmental issues within the corporate sector is that they compete with other activities or investment opportunities which are in practice more pressing. This is reinforced by the common response that a major external obstacle to environmental progress is the high cost of equipment. Clearly, the business perception of environmental protection is of a cost to be minimised.

Some of the other factors mentioned also point in the same direction, for instance, complaints of a lack of incentives, and possibly complaints about government policy. The former suggests that firms are, quite naturally, interested in environmental protection where financial incentives are in place for them to be adopted. The perception of government policy as an obstacle is more difficult to interpret. As Dominguez suggests this may simply reflect dissatisfaction on the part of the firms with certain government agencies, but cannot really be interpreted as a factor limiting firms' actions to control pollution.

Technological factors may also limit a firm's ability to tackle pollution problems. In the Mexican textile and iron and steel industries (although not in synthetic fibres) a number of firms felt that lack of technical information and of knowledge about environmental impacts and responses was an obstacle to environmental improvement. This suggests that a certain level of technological capability is required for efforts to reduce environmental damage to be undertaken. A similar conclusion can be drawn from the Argentine case study, which also highlighted the positive relationship between both active environmental management and pollution prevention and innovative capabilities (Tables 4.8 and 4.9). The links between technological capabilities and environmental improvements are illustrated in some detail in the case of the Brazilian pulp industry where, for instance, Aracruz's performance is directly linked to the innovatory capacity of the firm. Technological capabilities become increasingly important as firms seek to go beyond 'end-of-pipe' treatment and to introduce cleaner technologies.

Finance is also an important constraint. This means that firms which are in a weak financial position, and particularly small and medium enterprises with limited access to the capital markets, find it difficult to invest in environmental improvements. It also implies that industries that are in difficulties, such as the Brazilian tanning industry and the Mexican textile industry, find it particularly difficult to prioritise environmental issues. Furthermore, at times of economic crisis all firms may find it difficult to face additional environmental costs, and it has been observed that the authorities are well aware of this and may take a more relaxed approach towards enforcement of environmental regulations at such times, in order to avoid pushing firms over the edge.

The studies in this book provide very little evidence to indicate that environmental regulation leads not only to improved environmental performance but also to 'innovation offsets' which reduce costs (cf. Porter and van der Linde, 1995). This is not surprising since, as was pointed out above, most firms continue to rely on 'end-of-pipe' treatment rather than introducing cleaner technologies to address environmental problems.

This gloomy picture needs to be qualified slightly. As Chudnovsky *et al.* show, of the 70 important pollution prevention projects which they studied between 1992 and 1997, the cost of the project was recovered in a significant number, although generally the returns were lower than for non-environmental investments. Isolated examples of environmental improvements which generate economic benefits are cited in some of the other case studies as well, e.g. the reduction of waste in polymerisation and yarn making in one Mexican plant (see Chapter 9). There are also clear examples, as was seen earlier, where changes made to reduce costs have also reduced pollution. However, the general impression remains that attempts to reduce pollution and improve environmental performance lead to higher costs, and this is certainly how it is perceived by the majority of firms in Latin America.

Liberalisation and industrial pollution

The impact of the liberalisation of national economies on the environment has been a major area of debate in recent years. This issue was addressed in detail in Chapter 2 and discussed explicitly in several of the industry case studies. In Mexico and Argentina, particularly, the context within which firms have taken environmental decisions in recent years has been dominated by the opening of their economies. This is somewhat less true in the case of Brazil where liberalisation is more recent and less dramatic. Quite apart from the specific interest in the implications of liberalisation for the environment, it would be difficult to ignore such major economic changes in any discussion of industrial pollution.

As Chapter 2 has shown, however, there is no simple, universal relationship between liberalisation and pollution. The impact depends on a number of contingent factors including the structure of the economy, industry-specific factors, and national environmental policies and enforcement. These make it impossible to support generalisations that see liberalisation either as an unmitigated disaster for the environment, or in terms of a 'win–win' scenario.

Whereas Chapter 2 evaluates the overall effects of liberalisation, several of the other chapters provide insights into the impact at the firm and industry level and may pick up on aspects which more aggregate studies miss. Chapter 6 on the Brazilian leather industry is a case in point. Whereas most general studies regard leather as a relatively polluting industry, Odegard points out that most pollution is concentrated in a particular stage of the production process when 'wet–blue' is being produced. A major factor affecting the impact of liberalisation on pollution in the Brazilian leather industry is the shift in emphasis that has taken place within the sector towards 'wet–blue'. There is also some evidence of a similar shift towards exports of semi-finished products, which are relatively more polluting, in the Brazilian iron and steel industry (see Chapter 5).

This suggests that even fairly detailed aggregate studies, at the four-digit level of the International Standard Industrial Classification, may miss significant changes which are occurring within industries and have a major impact on pollution. It also supports the view that it is important to look at production in terms of 'commodity chains' in which different parts of the chain are distributed internationally with corresponding implications for the international distribution of pollution (see Knutsen, 2000).

At the firm level, some of the case studies suggest that production for export is an important factor in environmental performance. However, several studies also find that there is no direct relationship between the extent to which a firm produces for exports and its environmental performance. Nevertheless, in several cases firms which produce almost exclusively for the domestic market tend to be the worst performers. Does this mean that an increasing export orientation as a result of liberalisation leads to improved environmental performance? The problem here is that firms producing for the domestic market tend to be smaller, more technologically backward firms, so that it is not clear that production for export is of itself a critical factor. In other words it may be that firms which are best able to take advantage of the increased export opportunities associated with liberalisation are also those which are most able to improve their environmental performance, but that there is no direct causal relation between the two. On the other hand, as was seen above, in some industries, most notably the pulp industry, regulation in overseas markets has been critical in forcing exporters to improve their environmental performance. This is an area where further research would be valuable.

The studies provide limited evidence on the relative performance of foreign firms compared to those that are locally owned. They did not provide clear-cut evidence that foreign ownership was a key factor in determining performance, although there is evidence that foreign subsidiaries are disproportionately represented amongst the leading firms in terms of environmental performance (see the discussion of this point in Chapter 2). Even if it were the case that foreign firms tended to perform better than locally owned firms, this would not necessarily imply that increased foreign investment would lead to reduced pollution. For instance, in the case of the Brazilian leather industry, where foreign investment has been directed towards the most polluting production processes, it may well be that this has contributed to increased pollution.

Another aspect of liberalisation which is relevant to some of the industries studied in this volume is privatisation. All three countries have seen extensive privatisation in recent years affecting industries such as mining, petrochemicals and steel. The chapters which address this issue, particularly those of Warhurst and Hughes-Witcomb on mining and of Barton on iron and steel in Brazil, suggest that privatisation has been a positive factor. It is certainly true that, in the past, state firms were major polluters and that environmental regulators often treated them relatively leniently. Under these circumstances, if regulators are more willing to act against private corporations, privatisation is likely to have a positive effect.

Limitations and the need for further research

As was pointed out in Chapter 1, industrial pollution in Latin America has been relatively little researched and the studies in this book represent a first attempt at investigating some of the key issues in detail. Inevitably in such a situation, what can be achieved is limited, particularly when there is such a lack of available data. The case studies of key polluting industries have started to provide answers to some of the questions which were raised in Chapter 1, but it is important to be aware of the limitations of these studies.

In the absence of independent data, either because of a lack of monitoring or the fact that the authorities regard such information as confidential, all the studies rely heavily on information provided by the firms themselves. Given the sensitivity of environmental issues, there is a real question of how far the information provides a true picture of a firm's environmental performance. Respondents have an interest in providing as positive a picture as possible. There is also a problem of selection bias, in so far as firms who have provided access to the authors may well be those who take environmental issues most seriously.

A second problem is the tendency to focus more on policy and management, rather than on environmental impacts in discussing firm performance. This of course reflects the greater ease with which information on policy and management can be obtained, while information on impacts is often patchy and particularly difficult to obtain over time when firms have only recently begun to address such issues. As Chudnovsky *et al.* point out, environmental management does not necessarily lead to better performance in terms of environmental impact, although it is to be hoped that firms which develop environmental policies and introduce environmental management are in a better position to reduce pollution than those which do not.

A related problem arises where comparisons between firms are made, which rely mainly on indicators of policy and management. This may lead to biased results. For example, a large firm, or a multi-plant enterprise, may feel more of a need to adopt a corporate environmental policy or to introduce a standard environmental management system than would a smaller, single-plant firm. As a result, indicators which rely heavily on policies and management systems may suggest that large or multi-plant firms have a better environmental performance, whereas at the plant level, in terms of emission indices or recycling rates, the smaller firm might perform just as well, or better.

A further problem that arises when comparisons are made between different types of firms, is the difficulty of separating out different factors such as market orientation, ownership, size, etc., as determinants of environmental performance. Where research is based on a relatively small number of firms, as most of the case studies inevitably are, then it is impossible to separate out the impact of these different factors.

Many of these difficulties can be traced back to the lack of sufficient information about environmental impacts at the firm level. However, as environmental issues come to be taken more seriously by policy makers, and regulators come to adopt a more systematic approach towards pollution control, the information available to government will increase significantly. The challenge then is to ensure that this information is also available to the public and to researchers.

There is considerable scope for improving data collection and availability. It is quite common for a relatively small number of industries and firms to account for the bulk of pollution. Five out of a total of 27 three-digit International Standard Industrial Classification industries account for over 70 per cent of air emissions and toxic and metal pollution in Argentina, Brazil and Mexico.[10] Moreover, within each of these industries, pollution is often concentrated in particular sub-categories.

Equally it is often the case that a relatively small number of firms are responsible for the bulk of pollution in particular regions. For example, in the Mexico City metropolitan area, 94 firms account for more than 70 per cent of all atmospheric emissions (Poder Ejecutivo Federal, 1996: 103). In Mexico's second city, Guadalajara, 45 firms accounted for more than 80 per cent of all air pollution (Gobierno del Estado de Jalisco *et al.*, 1997: Table 7.12). A similar situation exists in the major Brazilian cities. In São Paulo, for example, 53 plants were estimated to account for 90 per cent of particulate emissions (Shaman, 1996: 6–7) while in Rio state the largest 100 sources were responsible for 70 per cent of all pollution (World Bank, 1996: Fig. 5). Thus, by concentrating on a relatively small number of major polluters and a few particularly polluting industries, it is possible to capture a large share of all emissions.

Even where data are presently collected, they are generally not readily available to the public or to researchers. There are, however, encouraging signs. The revised *Ley General del Equilibrio Ecológico y la Protección al Ambiente* in Mexico, approved in 1996, included a specific chapter on the right to environmental information, which could lead to much greater access, although I am not aware of any data at the firm level being released under this provision up to now. The government enforcement agency in Mexico, PROFEPA, is also drawing up an index of compliance with environmental regulation which it is proposed will be publicly available.

If and when such information does become available it will then be possible to look more systematically at the environmental impacts of firms and to link these to policy and management indicators. It will also become possible to look at a far larger number of plants or firms and carry out more sophisticated multivariate analysis of their performance.

Conclusion

The process of industrialisation has been underway for over a century in Argentina, Brazil and Mexico; however, it is only during the past decade that the environmental effects of this process have begun to be addressed. There is evidence in the studies in this book that some progress has been made during this latter period, but equally that these advances have been highly uneven and that major pollution problems still remain. Indeed, despite the evidence of progress, the overall impact of industry on the environment is towards increased levels of degradation.

A number of critical problems remain to be tackled. First, areas of rapid industrial growth have suffered major degradation because of the difficulties which local infrastructure and regulation have in keeping pace with increased emissions and waste. This was the case in Cubatão in the 1970s and is true of the Mexican border regions today (Barkin, 1999). Even if, at a national level, emissions were being brought down, the geographical distribution of pollution creates specific areas where large numbers of people suffer from increasing levels of pollution.

A second challenge is the need for business to go beyond the reactive response to government regulation, based on 'end-of-pipe' treatment measures which is the dominant pattern in Latin America today. There are examples of individual firms which have done this, so that the potential is obviously there, but so far these are the exceptions rather than the rule. This may well require new environmental policies, such as the use of market-based instruments, not as a substitute for regulation, but as an additional incentive for firms to go beyond compliance.

A third problem is to deal with the myriad of small and medium enterprises which exist in all three countries. However, although, as has been pointed out, these tend to pollute more than larger firms in relation to their output or employment, the bulk of pollution is caused by larger firms. Given also the difficulties of monitoring numerous small firms and, for many such firms, the cost of adopting pollution abatement measures, it makes sense to concentrate initially on larger companies. On the other hand, it is important to avoid creating a situation in which there is an incentive for large firms to outsource the most polluting production processes to SMEs which are less likely to be required to meet environmental standards.[11] One approach to this is to seek co-operation with large firms to ensure that suppliers comply with local environmental norms.

Industrial production in Latin America today is a long way from being sustainable. The advances noted in the studies in this book, although real, also illustrate how much there is still to do.

Notes

1 See Levy (1995) which found no relationship between the environmental practices of TNCs in the USA and their emissions of pollutants.
2 For evidence of a similar increase in energy efficiency in the Argentinian steel industry, see Chudnovsky *et al.* (1996: 507).
3 See below for a discussion of inter-industry differences in environmental performance.
4 Environment-related investment accounted for 6 per cent of total investment in the Brazilian iron and steel industry (see Table 5.7) compared to around 10 per cent amongst European steel firms (Barton, 1999: 147).
5 In Chapter 4, Chudnovsky *et al.* distinguished 'simpler' pollution prevention methods such as energy, water and input saving, good housekeeping, maintenance and operating practices and staff training, from more 'complex' measures such as process modifications, adoption of cleaner new technologies, raw material substitution and product reformulation.
6 This contrasts with the situation in the Mexican textile industry where the most common factor identified as the main source of environmental improvement was treatment.
7 The pulp and paper industry also affects the environment through its use of timber which has not been discussed here. For a much less favourable view of the environmental performance of the Brazilian pulp industry than that provided in Chapter 7, see Carrere (forthcoming) which focuses particularly on the ecological impact of the firm's timber supply policies.
8 A Mexican survey of firms in the chemical industry, food products, non-metallic minerals and metals found that over half the chemical firms had formal environmental management systems, compared to between 27 and 39 per cent in the other sectors (Lexington, 1996: IV-2).
9 Although this has been discussed as a response to market pressures and the demands of customers since this is how it appears directly to the exporting firm, it is of course also a response to regulation, albeit regulation in a country other than the one where the firm is producing.
10 The five industries are industrial chemical (351) petroleum refineries (353), other non-metallic minerals (369), iron and steel (371) and non-ferrous metals (372). This is based on estimates from applying the World Bank's Industrial Pollution Projection System (IPPS) coefficients to manufacturing value-added data for Argentina, Brazil and Mexico in the mid-1990s.
11 This has been termed 'source and hide pollution' by Bergstø *et al.* (1998).

Bibliography

AmCham (1993) *Environmental Compliance and Enforcement in Mexico: A Corporate View*. Mexico City: American Chamber of Mexico.

Barkin, D. (1999) *The Greening of Business in Mexico*. Geneva: United Nations Research Institute for Social Development, DP110.

Barton, J. (1999) 'Sectoral Restructuring and Environmental Management in the EU Iron and Steel Sector', *European Environment*, 9: 142–53.

Bergstø, B., S. Endresen and H. Knutsen (1998) '"Source and Hide Pollution." Industrial Organisation, Location and the Environment: Sourcing as a Firm Strategy', in H. Knutsen (ed.) *Internationalisation of Capital and the Opportunity to Pollute*. University of Oslo: FIL Working Papers, No. 14.

Carrere, R. (forthcoming) 'The Environmental and Social Effects of Corporate Environmentalism in the Brazilian Market Pulp Industry.' In P. Utting (ed.) *The Greening of Business in the South: Rhetoric, Practice and Prospects.*

Chudnovsky, D. and M. Chidiak (1995) *Competitividad y Medio Ambiente: Claros y Oscuros en la Industria Argentina*. Buenos Aires: CENIT, Documentos de Trabajo 17.

Chudnovsky, D., F. Porta, A. Lopez and M. Chidiak (1996) *Los Límites de la Apertura: Liberalización, reestructuración productiva y medio ambiente*. Buenos Aires: CENIT/Alianza Editorial.

Dominguez, L. (1999) 'Comportamiento empresarial hacia el medio ambiente: El caso de la industria manufacturera de la Zona Metropolitana de la Ciudad de México', in A. Mercado (ed.) *Instrumentos Económicos para un Comportamiento Empresarial Favorable al Ambiente en México*. Mexico City: El Colegio de México/Fondo de Cultura Económica.

Gobierno del Estado de Jalisco, Secretaría de Medio Ambiente, Recursos Naturales y Pesca and Secretaría de Salud (1997) *Programa para el Mejoramiento de la Calidad del Aire en la Zona Metropolitana de Guadalajara, 1997–2001*. Mexico City: Instituto Nacional de Ecología.

Knutsen, H. (2000) 'Environmental Practice in the Commodity Chain: The Dyestuff and Tanning Industries Compared'. *Review of International Political Economy*, 7(2): 254–288.

Knutsen, H. and K. Wik (1999) 'Tanning and the Environment in Mexico: the Case of León', in J. Hesselberg (ed.) *International Competitiveness: The Tanning Industry in Poland, the Czech Republic, Brazil and Mexico*. University of Oslo: FIL Working Papers, No. 15.

Levy, D. (1995) 'The Environmental Practices and Performance of Transnational Corporations', *Transnational Corporations*, 4 (1): 44–67.

Lexington (1996) *Industrial Environmental Management in Mexico: Report on a Survey*. Lexington: The Lexington Group, Report submitted to the World Bank.

Mello Lemos, M.C. de (1998) 'The Politics of Pollution Control in Brazil: State Actors and Social Movements Cleaning up Cubatão', *World Development*, 26 (1): 75–87.

Poder Ejecutivo Federal (1996) *Programa de Medio Ambiente 1995–2000*. Mexico City: Instituto Nacional de Ecología.

Porter, M. and C. van der Linde (1995) 'Toward a New Conception of the Environment–Competitiveness Relationship', *Journal of Economic Perspectives*, 9 (4): 97–118, Fall.

Roberts, T. and E. Stauffer (2000) *Corporate Environmentalism in Brazil's Chemical Industry: Participation in Responsible Care and ISO 14001 in 619 firms*, paper for presentation at the Latin American Studies Association XXI International Conference, Miami, FL, 16–18 March.

Runge, C. F., F. Cap, P. Faeth, P. McGinnis, D. Popageorgiou, J. Tobey and R. Housman (1997) *Sustainable Trade Expansion in Latin America and the Caribbean: Analysis and Assessment*. Washington, DC: World Resources Institute.

Scholz, I., K. Block, K. Feil, M. Krause, K. Nakonz and C. Oberle (1994) *Medio Ambiente y Competitividad: el Caso del Sector Exportador Chileno*. Berlin: Instituto Alemán de Desarrollo, Estudios e Informes 13/1994.

Shaman, D. (1996) *Brazil's Pollution Regulatory Structure and Background*. Washington, DC: World Bank.

World Bank (1996) *Brazil: Managing Environmental Pollution in the State of Rio de Janeiro*. Vol.1. *Policy Report*, Washington, DC: World Bank.

INDEX

For Product Safety Concerns and Information please contact our EU
representative GPSR@taylorandfrancis.com
Taylor & Francis Verlag GmbH, Kaufingerstraße 24, 80331 München, Germany

www.ingramcontent.com/pod-product-compliance
Ingram Content Group UK Ltd.
Pitfield, Milton Keynes, MK11 3LW, UK
UKHW021117180425
457613UK00005B/130

* 9 7 8 0 4 1 5 2 3 4 4 7 4 *